Professional Developer's Guide

WAP Servlets

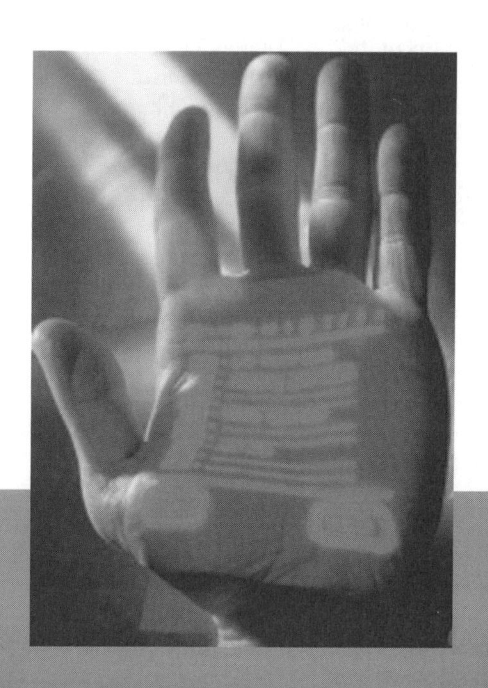

John L. Cook, III

Wiley Computer Publishing

John Wiley & Sons, Inc.

NEW YORK • CHICHESTER • WEINHEIM • BRISBANE • SINGAPORE • TORONTO

Publisher: Robert Ipsen
Editor: Carol Long
Associate Editor: Margaret Hendrey
Managing Editor: Micheline Frederick
Associate New Media Editor: Brian Snapp
Text Design & Composition: D&G Limited, LLC

This book is printed on acid-free paper. 

Published by John Wiley & Sons, Inc.

Published simultaneously in Canada.

Library of Congress Cataloging-in-Publication Data:

ISBN: 0-471-39307-X

Printed in the United States of America.

10 9 8 7 6 5 4 3 2 1

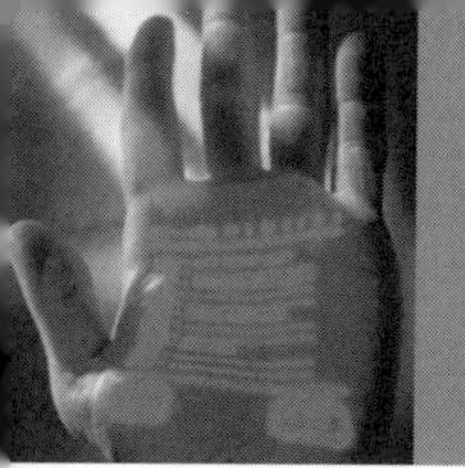

Professional Developer's Guide Series

Other titles in the series:

Advanced Palm Programming by Steve Mann and Ray Rischpater, ISBN 0-471-39087-9

Java™ 2 Micro Edition by Eric Giguère, ISBN 0-471-39065-8

To my daughter Rachel.

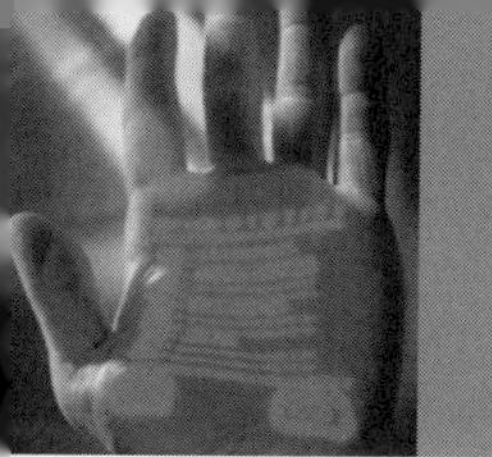

Contents

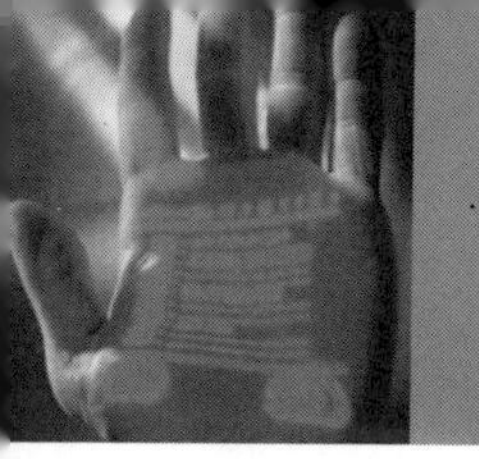

Acknowledgments

This book would not have been possible without the help of the following people:

- First and foremost, I would like to thank my wife Kathy and daughter Rachel for their patience and understanding—especially for putting up with the crazy hours. I should also thank our cats Atticus and Boo for keeping me company during those late nights.
- Thanks to Carol Long (Executive Editor) and the other fine folks at John Wiley & Sons, Inc.
- Thanks to my friends, the Snows, for providing an excellent venue for the writing of Chapter 5.
- Thanks to the people of Sun Microsystems for providing the JSDK to simplify Web application development for the masses.
- Thanks to the various members of the WAP Forum for their past, present, and future efforts in developing a protocol suite for the wireless Internet.
- Thanks to Mom and Dad for putting me through school and giving me the support and encouragement to pursue my occupation of software engineer.

To all of you, my most sincere appreciation.

John L. Cook III

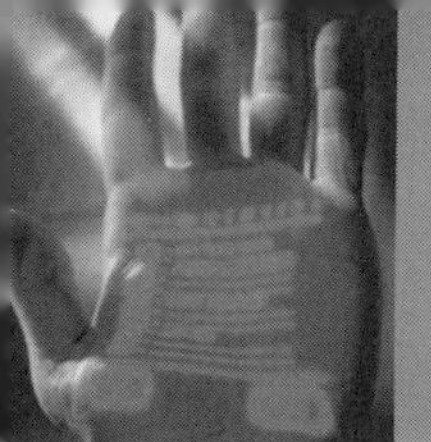

CHAPTER 1

Introduction

The Internet has changed the way individuals work, play, and live. The advent of the Internet is the single most important happening in computing history and has changed the desktop computer from a business-productivity tool to a household appliance. The ultimate impact that the Internet will have on humanity is just beginning to unfold. In the early stages of this revolution in technology, a relatively new twist is beginning to have its own profound influence: the wireless Internet. The primary vehicle for the transition from wired to wireless Internet is the hand-held browser, and the developer's key to unlocking its mysteries is the WAP Servlet.

Why Develop for the WAP Browser?

Perhaps the first question that you should ask is, "Why develop for the wireless Internet?" After all, Internet companies are coming out with unbelievable evaluations, and these companies do not necessarily have anything to do with wireless technologies. Furthermore, many of these companies do not even have a wireless strategy. Developers would certainly not be making a mistake by focusing on traditional wired networking.

In order to see why the wireless Internet is an equally intriguing choice, you should consider two areas. First, how has the wired Internet impacted society? Second, how does wireless technology augment and improve the wired Internet? By examining these topics, you should see that wireless technology will have a profound effect on how people use the Internet in the future.

So how exactly has the wired Internet impacted humanity? You might begin to answer this question by reviewing your own online shopping habits over the last few years.

Online shopping (or e-commerce, as it is commonly known) has many advantages to the consumer: no travel time, more product information, better prices, and delivery in the case of gift shopping. Besides the advantages to consumers, retailers also gain benefits from e-commerce: a new channel, reduction in sales staff, and lower facility costs.

The Internet has also improved individuals' abilities to communicate. Workers can choose from many technologies that improve their ability to communicate with coworkers, customers, and partners: e-mail, conferencing, Web sites, and online tools. Virtual private networking has enabled individuals to participate in their corporate networks regardless of their location. Web-based communications services have enabled people to access their e-mail and voice mail from virtually anywhere on the planet. Additionally, you cannot forget the roots of the Web in its ability to convey information. Using this unique aspect of the Internet, corporations and individuals have the ability to make information known to every Internet user in a matter of moments.

Besides its utility for e-commerce and business at large, the Internet has rapidly moved into the realm of entertainment. The Internet has extended the value of existing media (such as radio, television, periodicals, and books) by supporting its content with online supplements. New Web-based streaming audio and video programs are on the market. Both individual and multi-player games are available for the entertainment of Internet users. With all that the Internet has brought to humanity, it does have one fundamental limitation, however: portability. In order words, e-mail, e-commerce, gaming, streaming video, surfing, and so on all require sitting at a desktop system—either at home or at the office. The ability to connect a laptop via a cellular phone or wireless modem has greatly improved portable access to the Internet; however, there are many situations when a wireless laptop is inconvenient or unavailable. Furthermore, when moving from location to location, you tend to turn off these devices and pack them in cases that are not convenient to carry in many situations.

New wireless hand-top and palm-top devices (such as the Palm OS and Windows-CE devices that have been wirelessly enabled) have been introduced in recent years. While these devices improve the ability to be connected while away from the desktop, they are unable to provide the same level of experience that users have come to expect from the Internet. Users are beginning to realize, however, that the goal of the wireless Internet is to augment the wired Internet, not to replace it.

Technology has extended the reach of the Internet past the wire. The *Wireless Application Protocol* (WAP) standards define the framework for connecting a WAP browser to the Internet. The WAP browser is a software application that is suitable for implementation in common hand-held wireless devices: cellular phones, two-way pagers, and personal information managers. Unlike a laptop or palm-top that uses standard Internet protocols over a wireless modem, the WAP architecture was designed to support rather than replace the wired Internet. Because the WAP browser runs on devices that are designed to be carried about and operated around the clock, WAP leads to an important evolution in Internet usage: 24-hours-a-day, seven-days-a-week connectivity.

Full time connectivity to the Internet means that you can stay connected with your desktop e-mail, contact list, and calendar. This functionality also means that you can surf the Internet, read the news, or play a game while standing in line at the grocery store. You can also compare prices and order a product via e-commerce while looking at that product in a showroom. The bottom line is that we have a paradigm shift.

Why develop applications for WAP browsers? These applications are the next wave of the Internet.

Why Read This Book?

This book brings together two essential components required for modern WAP development: Servlets and the *Wireless Markup Language* (WML). Sun Microsystems *Java Servlet Development Kit* (JSDK) provides a simple framework for generating dynamic Web content. The WAP Forums WML and WML Script (a scripting language for WAP browsers) provides the content for *WAP clients* (hand-held devices). Together, these two technologies provide developers with the tools required for WAP development. Furthermore, WAP and Java are a logical pairing because they are both new, Web-centric development tools.

What Is in This Book?

This book contains eight chapters and four appendices. The following paragraphs describe the contents.

Chapter 1, "Introduction," provides an introduction to the book and its content.

Chapter 2, "Preliminaries," provides basic information about the technologies explored in the book. In this chapter, we define the *Hypertext Transfer Protocol* (HTTP) client and server. We also present elements of the WAP architecture.

Chapter 3, "Wireless Markup Language (WML)," looks at the essentials of XML. There, we define the *Document Template Definition* (DTD) for WML. We describe the various elements and attributes of the WML DTD in a tutorial format. As we define the elements and attributes, we show detailed examples and a sample screenshot.

Chapter 4, "Java Servlet Development Kit," explores the interfaces and classes of Sun's JSDK. In particular, we explore the most essential methods of five JSDK classes/interfaces. By limiting our discussion to those methods that are most useful to the WAP Servlet developer, this chapter provides a focused look at a power facility for developing Web content in a tutorial format.

Chapter 5, "WML Home Page Example," brings together the technologies discussed in Chapters 3 and 4 with a real-world example that renders a WML homepage.

Chapter 6, "Grocery Shopping List Example," builds upon the previous chapters by providing an application-oriented example of a WAP Servlet. Rather than serving or rendering static pages (as the examples in Chapter 5 show), this example dynamic application renders responses based on user actions.

Chapter 7, "Push Technology," provides an overview of the WAP push architectures and the *Push Access Protocol* (PAP).

Chapter 8, "WML Script," provides a basic overview of WML Script. This scripting language provides developers with the capability to imbed computational software within a WAP browser. The overview of WML Script is supported by a simple example: Appendix A, "WML 1.2 Reference."

Appendix A, "Wireless Markup Language (WML) Reference," supports the needs of WAP developers once they have completed the tutorial presented in Chapter 3. This chapter provides a quick reference to the elements and attributes of the WML 1.2 specification.

Appendix B, "JSDK 2.1.1 Reference," supports the needs of WAP developers once they have completed the tutorial presented in Chapter 4. This chapter describes all of the classes and methods of the 2.1.1 JSDK in reference format.

Appendix C, "HTTP Servlet Server," presents a source code listing of the Servlet Engine supplied with the book on the enclosed CD-ROM. While neither a complete implementation nor a commercial-grade product, the Servlet Engine is suitable for hosting the examples in this book and will support the develop needs of most WAP developers. This chapter shows how HTTP servers utilize the interfaces provided with the JSDK in order to implement Servlet engines.

Appendix D, "Companion CD-ROM Contents," provides information about the contents of the CD-ROM.

Expected Background

Because this book deals with writing WAP Servlets in Java, the user should have prior experience developing in Java. While someone who is knowledgeable in C++ or another object-oriented language should be able to understand most of the example code, it is certainly not a goal of this book to train the reader in Java development. While not required, prior experience with *HyperText Markup Language* (HTML) will make understanding WML much easier. Likewise, experience with Java Script will assist readers with understanding WML Script. Prior experience with JSDK, WML, or HTTP is not required.

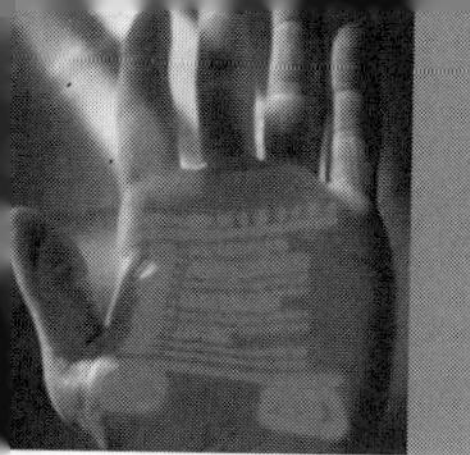

CHAPTER 2

Preliminaries

The Internet is a complex fabric of computers and communications equipment controlled by equally complex networking software. The wireless Internet adds to this mix by introducing new hardware and software components. The Java Servlet pulls everything together from an application point of view. In this chapter, we will examine the building blocks of the Internet and the background knowledge required for developing WAP Servlets.

The Wired Internet

Oddly enough, no discussion of the Internet would be complete without at least briefly mentioning the *International Standards Organization Open Systems Interconnect* (ISO/OSI) model. The irony is that more than a decade ago, people predicted that the ISO networking standards would replace the most popular protocols used on the Internet today: *Transmission Control Protocol/Internet Protocol* (TCP/IP). Not only has this result not happened, but TCP/IP has practically become a household term. Yet, it is still popular in communications circles to refer to the ISO/OSI model when discussing networking protocols. To that end, Figure 2.1 shows a mapping between the seven layers of the ISO/OSI model and the protocols commonly used on the Web.

The seven layers of the ISO/OSI model are as follows: Hardware, Data Link, Network, Transport, Session, Presentation, and Application. The following sections look at the Internet protocols that map onto the model. By looking at these protocols, we can gain additional insight into Web application development.

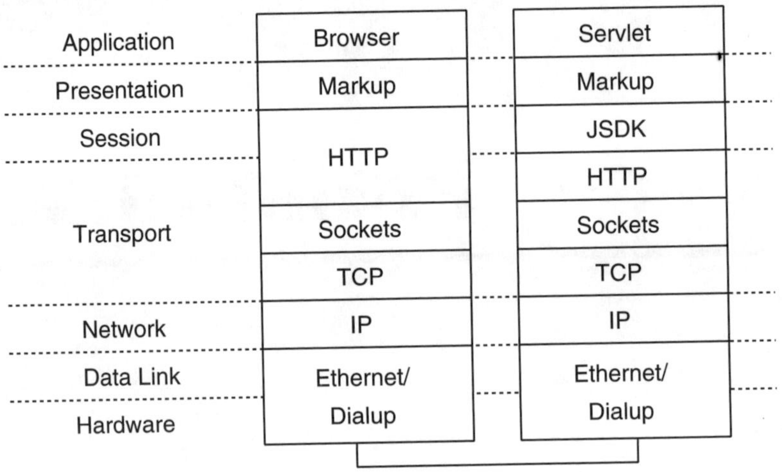

Figure 2.1 ISO/OSI view of the Internet.

Hardware/Data Link

While Al Gore might not have invented the Internet, as far as I know, he is responsible for the phrase "the information superhighway." This analogy is actually extremely good. Consider the road system in the United States. Essentially, all houses are connected to the road system and are therefore connected to each other. In fact, given any two houses on this road system, you can drive from one to the other with a surprisingly few number of turns. Likewise, most computers have some connection to the Internet, and when a system is connected to the Internet, it is connected to every other machine on the Internet. Furthermore, there are a surprisingly few number of machines through which traffic between two machines must flow. Hence, we have a networking *superhighway*.

While not paved with asphalt, the roads of the Internet are constructed with hardware and data link protocols. The Hardware layer primarily deals with the modulation of data as tones or electrical impulses that are suitable for transportation over a wire, light over fiber, or radio waves though the air. The Data Link layer deals with microcode and device drivers that are necessary for the Hardware layer. Together, these layers provide the virtual road system that connects all computers.

The Network Layer

While the *Transmission Control Protocol* (TCP) and the *Internet Protocol* (IP) are typically joined at the hip with the ever-so-common acronym TCP/IP, they are in fact two extremely distinct protocols. IP resides at the Network layer. The primary function of IP is to encapsulate and deliver a portion of data between two networked computers. This portion of data is referred to as a packet. A side effect of IP is that it hides the implementation details of both the Data Link layer and the hardware used. Refer to the Internet Protocol RFC [RFC 791] for additional information.

NOTE

The IP protocol initially made its debut in the Department of Defense ARPANET (Advance Research Projects Agency Network). The ARPANET was created in response to the Soviet Union launching its Sputnik satellite. The United States felt that it was falling behind the Soviet Union in high technology pursuits. During the 1980s, as the public sector became more interested in networking technology, the ARPANET transitioned into the Internet. The then-popular network layer protocol of the ARPANET, IP, became the Network layer of the Internet.

When you send a letter via the postal service, you enclose your letter in an envelope. The envelope itself contains important information used to deliver its contents (the recipient's name and the return address). You can think of the IP packet as a sort of electronic letter. The packet consists of a header and payload. The header, much like an envelope, contains addressing information about the source and destination. The payload of a packet is simply the data portion of the message. Continuing with the postal service analogy, both the sender and the recipient are computers, and the postal service is the Internet. Figure 2.2 shows an IP packet consisting of a header and payload section.

TCP

While the IP protocol provides powerful functionality by itself, that functionality is limited. Without using additional protocols, you could deliver packets from one machine to another; however, it would be awkward to deliver that data to an application. Of course, we could include the name of the application or some identifier within the payload of the packet. (In the early days of communications, this procedure is exactly what was

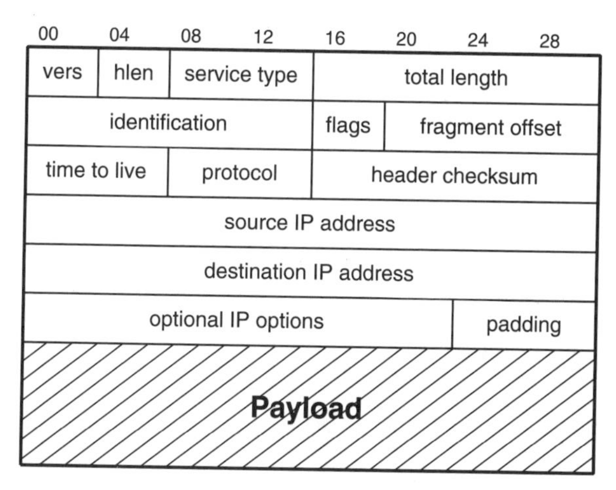

Figure 2.2 IP packet.

done.) The Internet has two protocols that operate above IP to provide this functionality, however: TCP and the *User Datagram Protocol* (UDP). Refer to the UDP [RFC 768] and TCP [RFC 7930] RFCs for additional information.

The UDP protocol is the lesser-known sibling of TCP. Because this topic is not germane to the subject of this book, I will not discuss it at length; however, there are a couple of things that we should discuss in order to contrast UDP with TCP. The UDP protocol delivers a datagram. A datagram is a single packet that represents a complete message by itself. When using UDP, you have no guarantee that the packets will be delivered in order. In fact, there is no guarantee that packets are delivered at all. Furthermore, due to the complexities of network routing, it is possible that a datagram might be delivered twice. When you compare this limited functionality to the functionality of TCP, I am sure that you will understand why the phrase "UDP/IP" has not gained the same level of popularity as TCP/IP.

Despite its limitations, the UDP protocol does provide a means for identifying which application sent and which application is to receive the datagram. This identifier is a two-byte value referred to by the term *port*. The UDP header contains two ports: the *source port* and the *destination port*. As you might guess, the source port identifies the sending application, and the destination port indicates the desired recipient application. Besides the header, a UDP packet contains payload. The UDP header and its payload are always delivered as the payload of an IP packet. Figure 2.3 shows the organization of a UDP datagram.

With that brief discussion of UDP behind us, we are ready to delve into the realm of TCP. Like UDP, TCP contains its own domain of two-byte ports. Its header contains a source and destination port to identify the source application and the desired recipient application, respectively. Unlike UDP, TCP provides the notion of a connection. A TCP connection is a bidirectional path between two applications over which you can send a continuous stream of information. The stream of data is broken into the payload section of multiple TCP packets. Each TCP packet is delivered as the payload of an IP packet. As the destination machine receives TCP packets, the individual TCP payloads are reassembled into a stream for presentation to the receiving application.

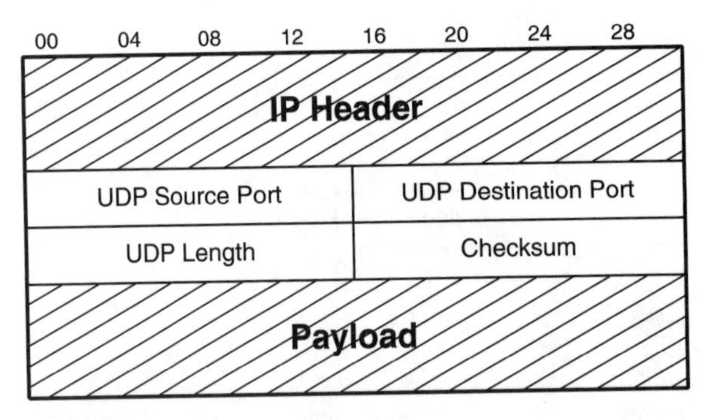

Figure 2.3 UDP datagram.

Because a stream of data is broken into multiple packets, TCP must handle three basic elements of data integrity. First, TCP must eliminate duplicate packets. Second, out-of-order packets must be reordered before the stream is reassembled. Three, TCP must recover lost packets. Fortunately, TCP handles all three of these elements without any additional effort expended by the applications on either system.

NOTE

The TCP protocol does not guarantee data lost due to a lost connection; however, while a connection is open, it does guarantee the integrity of the stream. Lost packets are recovered, and duplicate packets are eliminated. Out-of-order packets are reordered. The UDP protocol provides none of these data integrity features.

At this point, I would like to mention one additional feature of TCP. Today's Internet applications place a great deal of data into TCP streams. Graphics-rich Web pages, streaming video, and *Voice over IP* (VoIP) applications served by mega Web servers could literally drown a smaller desktop machine with data. The TCP protocol implements an algorithm known as a sliding window. The *sliding window* enables a machine to announce the amount of memory available for receiving data from the other machine. This functionality enables a TCP connection to be throttled.

Figure 2.4 shows the organization of a TCP packet.

Via Sockets

Together, a TCP port and an IP address are referred to as a socket. This pairing lent its name to a variety of libraries that enable you to send datagrams (typically via UDP) and

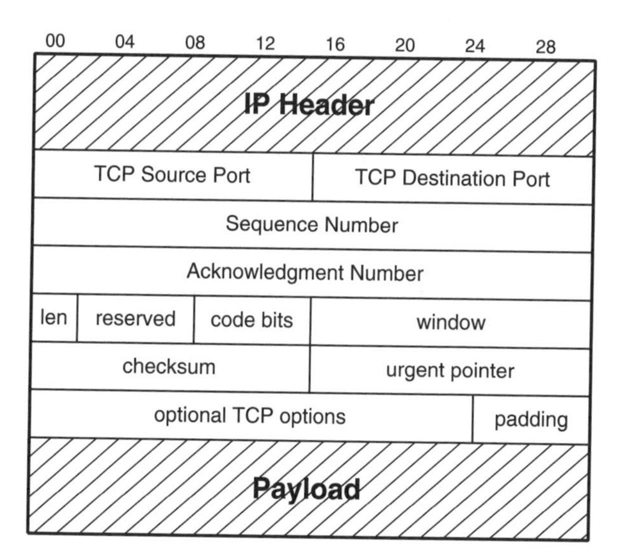

Figure 2.4 TCP packet.

streams (usually via TCP). Within the Java world, there is an answer to the socket library function contained within the `java.net` package. Of primary importance to Java TCP developers are two classes: `java.net.Socket` and `java.net.ServerSocket`. The `java.net.ServerSocket` class enables a server to claim a port for all incoming connection requests. As connections are established, the server is provided with a `java.net.Socket` instance that represents that connection. The `java.net.Socket` class provides a means for clients to connect to a server port. The `Socket` class provides two methods of interest: `getInputStream()` and `getOutputStream()`. As you might expect, these methods give you access to a `java.io.InputStream` for reading streamed data and a `java.io.OutputStream` for writing streamed data.

While I showed Sockets as a layer in the ISO/OSI model (refer to Figure 2.1), it does not physically change the data sent between two machines (as in a protocol header); however, it is part of the path that the data follows between the two machines. For example, when a client writes data to the socket's OutputStream, it arrives on the server's InputStream. The Java implementation of sockets is elegant in its simplicity. For example, Figure 2.5 shows a complete implementation (albeit simple) of a Java socket-based client and server.

Hypertext Transfer Protocol (HTTP)

In the early days of the Internet, users used several discrete protocols. At that time, it was popular to have a separate protocol for each function. For example, there were protocols for e-mail (Simple Mail Transfer Protocol [SMTP]), for file transfer (File Transfer Protocol [FTP], and Trivial File Transfer Protocol [TFTP]), and for remote access (Telnet). As network technology matured, general-purpose facilities developed to enable client/server networking in traditionally stand-alone applications (such as remote procedure calls). In their infancy, network applications were rudimentary; however, as desktop computers became popular and software became more sophisticated, network applications evolved. An important step in this evolution was applications that enabled users to point and click their way through network documents.

The predecessor to the Internet browser was the File Gopher [RFC 1436]. By using Gopher, users were able to access files on a given server by using a *Graphical User Interface* (GUI) much like they would access files on their own system. Around the same time that Gopher was working its Internet magic, a new encoding method was invented to mix media in mail messages: *Multi-Purpose Internet Mail Extensions* (MIME) [RFC 2045]. Both of these technologies lead to the *Hypertext Transfer Protocol* (HTTP) [RFC 2616].

HTTP Application Framework

While many machines are involved in network communications at large, we should discuss three machines relative to HTTP: *clients*, *servers*, and *proxies*. In Internet terms, a server is any machine that binds a TCP port for the purpose of providing a service to would-be clients. The client is any machine that connects to a well-known port on a server in order to

```
package book.chap02.HelloWorld;

import java.io.*;
import java.net.*;

public class Main extends Thread
{
  public static void main(String strArgs[])
  throws Exception
  {
    new Main().start();
    Thread.yield();

    Socket socket = new Socket("localhost", 2000);
    OutputStream os = socket.getOutputStream();
    OutputStreamWriter osw = new OutputStreamWriter(os);
    BufferedWriter bw = new BufferedWriter(osw);
    PrintWriter pw = new PrintWriter(bw);

    pw.println("Hello, world");
    pw.close();
  }

  public void run()
  {
    try
    {
      ServerSocket serverSocket = new ServerSocket(2000);
      Socket socket = serverSocket.accept();

      InputStream is = socket.getInputStream();
      InputStreamReader isr = new InputStreamReader(is);
      BufferedReader br = new BufferedReader(isr);

      System.out.println(br.readLine());
    }
    catch (IOException e)
    {
    }
  }
```

Figure 2.5 A simple demonstration of Java sockets.

access a service. A proxy is a machine that channels client connections on a *Local Area Network* (LAN) into the Internet. The primary function of a proxy is to facilitate dialing to an *Internet Service Provider* (ISP). A secondary function of a proxy is to monitor which

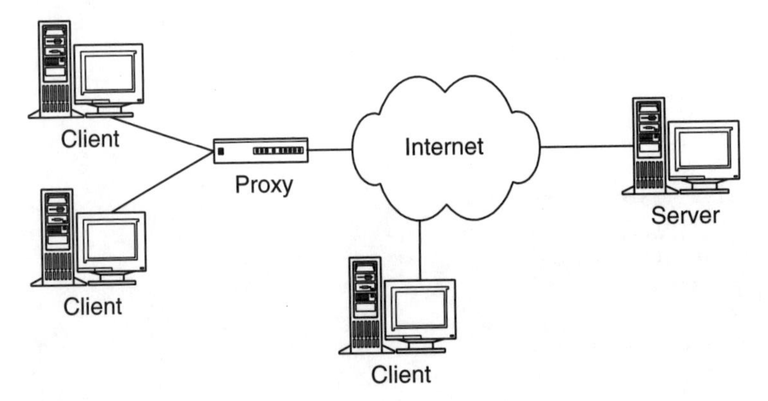

Figure 2.6 Clients, servers, and proxies.

requests that end users are making on the Internet. Figure 2.6 shows three clients accessing a server.

NOTE

Due to recent improvements in operating systems and the proliferation of firewalls, proxies are not as popular as they were at one time. Many firewalls have proxy functions. Because home users have traditionally called an ISP from their desktops, a proxy does not really make much sense. A bit of a proxy revival is taking place, however, as users (especially of broadband and *Digital Subscriber Lines* [DSL] ISPs) are setting up LANs in their homes. There are also hybrid products (proxy and firewall) that target this market (for example, Ositis' WinProxy).

Besides the HTTP proxy software, there are two chief applications that utilize the HTTP protocol: browsers and the HTTP daemon. (A *daemon* is an application that provides a network service on a server. Many times, daemons are referred to as servers.) The daemon binds one or more TCP ports for incoming HTTP requests. The TCP port 80 is the typical port used for non-secure HTTP service. The TCP port 443 is used for secure communications.

The main function of an HTTP browser is to make a request for document resources on the Internet. A *Uniform Resource Locator* (URL) is used to identify a desired resource. Because a URL consists of a server name and a *Uniform Resource Identifier* (URI), the browser can open a connection to the desired daemon and make a request for the document identified by the URI.

NOTE

A URL consists of four parts: a scheme, a server *Domain Name Service* (DNS) name, a TCP port, and a URI. The format of a URL is as follows: `scheme://dns-name:tcp-port/URI`. Most browsers will assume a scheme of `http`. The TCP port is optional and will default on the server to the daemon associated with the scheme (80 for HTTP and 443 for HTTPS). The URI is also optional.

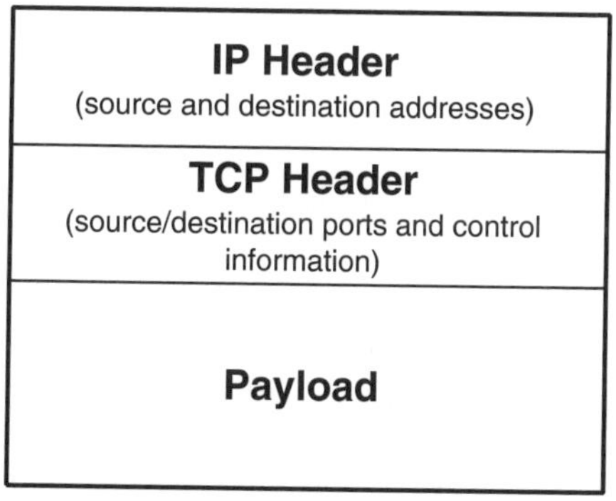

IP Header
(source and destination addresses)

TCP Header
(source/destination ports and control information)

Payload

Figure 2.7 HTTP packet.

HTTP Requests and Responses

The HTTP protocol operates over TCP. Figure 2.7 shows the organization of an HTTP packet. Just like TCP and IP, HTTP consists of a header section and a payload section. HTTP is a synchronous request/ response protocol. In other words, for each request, there is one (and only one) response. Therefore, there are two types of HTTP packets: requests and responses.

Unlike TCP and IP, which both have binary headers, the HTTP header consists of printable ASCII. Each line of the header is terminated by a carriage return/line feed (0x0D0A). The header itself is terminated by an additional carriage return/line feed. The content portion of the HTTP packet can be ASCII or binary, as specified in the header option `content-type`. Either the end of the TCP stream or the value specified in the header option `content-size` determines the size of the content.

HTTP Header

The first line of an HTTP header differs between the request and the response. For a request, this line consists of three arguments: `method`, `URI`, and `version`. The `method` determines the type of request (and, for the purposes of this book, can be either *get* or *post*). The `URI` identifies a resource on the server. The `version` informs the server which version of the HTTP protocol that the browser supports. Each of these fields is delimited by a single space. The first line of a response also consists of three fields: `version`, `error code`, and `error message`. As with the request, these fields are delimited by a single space; however, the error message can contain spaces. The `version` enables the server to inform the client of which version of the HTTP protocol that it supports. The `error code` is a three-digit decimal number that indicates any errors that occurred in processing the request. This value is machine-readable. The `error message` is a human-readable message that describes the `error code`.

The balance of lines in an HTTP header consists of *option/value pairs* (OVP). Each OVP consists of an option name and value, delimited by a colon. The value of the `content-type` option is the *Multi-Purpose Internet Mail Extensions* (MIME) type of the content.

The value of the `content-size` option is the length of the content in bytes. Other OVPs are typically exchanged between the browser and server in order to negotiate and/or describe the requests and responses.

Markup Language

Prior to the Internet, markup languages were primarily used in word-processing applications. Markup enables a user to format his or her text documents for improved readability by using tags. Bar none, the most popular markup language on the Internet is *Hypertext Markup Language* (HTML). While a discussion of HTML is beyond the scope of this book, a brief discussion is in order. Refer to the Hypertext Markup Language RFC [RFC 1866] for more information.

NOTE

HTML has a structure that is similar, yet more loose than the Extensible Markup Language (XML). We discuss the XML language in Chapter 4, "Java Servlet Development Kit."

HTML uses elements, and elements typically have a start tag and an end tag that offset the text that they influence. An HTML document starts with an `<HTML>` tag and ends with an `</HTML>` tag. The visible section of a document resides within the `<BODY>` and `</BODY>` tag pair. A paragraph is offset by `<P>` and `</P>` tags. Figure 2.8 shows a simple example of an HTML document.

The HTML language has many more elements than we discuss here. Most of the elements take attributes that further customize their impact on the text that they enclose. Using the full capabilities of HTML enables users to create content-rich Web pages. Figure 2.9 shows the organization of an HTML message. This figure shows a single TCP/IP packet. If the payload (markup) portion of an HTTP response were large enough, it would be broken up over several TCP/IP packets.

A major function of Internet markup languages is their capability to link documents. This feature clearly differentiates the HTTP/HTML pairing from Gopher, and in my humble opinion, this functionality marked the genesis of the modern Internet.

```
<HTML>
<BODY>
<P>This is the first paragraph.</P>
<P>This is the second paragraph.</P>
</BODY>
</HTML>
```

Figure 2.8 HTML example.

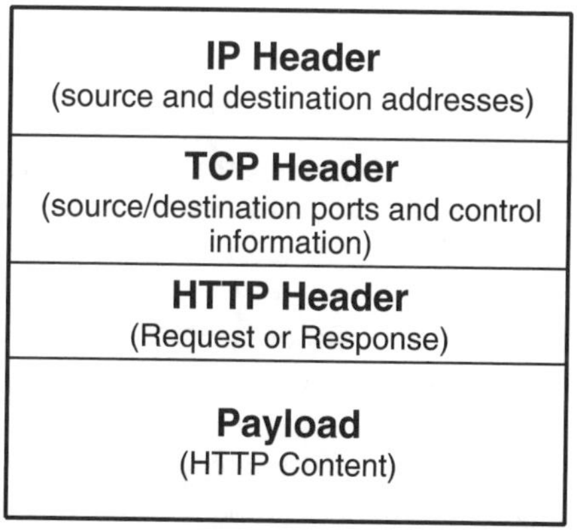

IP Header
(source and destination addresses)

TCP Header
(source/destination ports and control information)

HTTP Header
(Request or Response)

Payload
(HTTP Content)

Figure 2.9 HTML message.

NOTE

The term *surfing* comes from the ability to click on a link in one document and load another document. Considering that the search engines on the Internet are merely rendering dynamic HTML documents, in effect, you can access most documents on the Internet by clicking a link in another document.

Web-Based Java Applications

While the simple ability to access documents anywhere in the world as if they reside on your computer is both fascinating and powerful, it is also limited. The new Internet economy is more a function of Web-based applications than document surfing. Many technologies are available to the *Webmaster* (a person who is responsible for an Internet site): Microsoft's *Active Server Pages* (ASP) and Sun's *Java Server Pages* (JSP) and *Enterprise Java Beans* (EJB). The technology that we will focus on in this book, however, is Sun's Java Servlets.

All of the application extensions mentioned earlier, as well as many others, owe their origin to an aspect of HTTP servers: the *Common Gateway Interface* (CGI) script. CGI scripts are often referred to as CGIBIN scripts, because they tend to reside in a directory called CGIBIN. By using CGI, Webmasters are able to associate shell scripts and/or applications with URIs on their server.

A URI actually consists of two parts: a virtual path and a query string. These two components are separated by a question mark (?). The optional query string is made up of one or more name/value pairs delimited by an ampersand (&). The equals sign is used to delimit the name and value parts of a name/value pair. The following string demonstrates the format of a URI: `/path?name1=value1&name2=value2`.

If the path of a URI identifies a static document on the server, the query string portion is ignored. If a URI identifies a CGI script, however, the query string represents arguments that are passed to the script. By using the CGI capability, an HTML document (for example) can contain a link to a script and pass it arguments. For example, you can implement a shopping cart by using a scripting language such as Perl.

While CGI greatly extended the capabilities of HTTP servers, there is a significant price to pay. Each time a CGI script is invoked, the HTTP server must create a process, perform a context switch to that process, load a script application, execute the script, context switch back to the server, and clean up the process. A user of a lightly loaded server might get the impression that these steps are processed rapidly; however, on a heavily loaded server, this inefficient and costly method will quickly disappoint users because it will become the processing bottleneck. Considering that it is often necessary to maintain session information (such as shopping cart contents for an online order) across multiple invocations of a script, many CGI scripts must save data to system files. Obviously, relatively slow disk *input/output* (I/O) will further impact performance.

Due to its powerful suite of networking classes, Java has quickly become a popular language for implementing HTTP servers. To carry CGI to the next level, Sun defined the *Java Servlet Development Kit* (JSDK) [JSDK 2.0]. By using JSDK, developers can develop CGI-like functionality in the form of a Java Servlet. The primary advantage of this function is that Servlet invocation executes within a thread, rather than within a process. Because Java threads are much more efficient than system-level processes, the JSDK scales much better than CGI. Also, considering that Servlet session information (such as a shopping cart's contents) can reside in memory between requests, there is a vast improvement in the overhead associated with data persistence.

NOTE

The JSDK is primarily a collection of Java interfaces. HTTP server developers provide implementation classes behind these interfaces to enable JSDK Servlets to execute from their server.

The Wireless Internet

So, you can implement static Internet documents. You can link your documents to documents on other servers, and vice-versa. You can implement JSDK Servlets that supplement your site with applications. Your server is running on the Internet 24 hours a day, seven days a week, and anyone in the world who has a desktop computer and an ISP can access your content. Life is good, but something is missing: content for users who have smaller, hand-held devices.

The WAP Browser

The *Wireless Application Protocol* (WAP) architecture extends the wired Internet model for wireless, hand-held browsers. From a hardware perspective, we must add an additional element in order to handle HTTP requests and responses within the WAP domain: the WAP

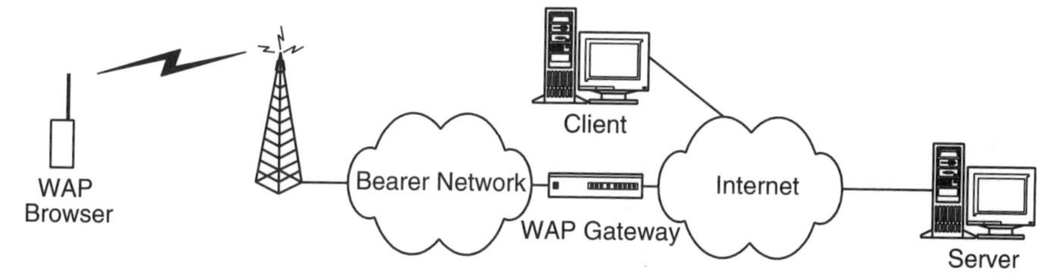

Figure 2.10 WAP hardware architecture.

gateway. The WAP gateway is responsible for interfacing the wireless bearer network to the wired Internet. Figure 2.10 shows the WAP hardware architecture.

NOTE

One of the reasons for the existence of the WAP gateway is to enable WAP browsers to access standard HTTP servers. From a Webmaster's point of view, this function is essential. Rather than having to build a new server infrastructure or leasing such technology from a third party, an HTTP server operator can extend sites reach to the wireless world by merely generating a new type of markup served by the same server that he or she currently uses.

From an HTTP perspective, the WAP browser is similar to other HTTP browsers; however, the primary content of interest to a WAP browser is the *Wireless Markup Language* (WML) and WML Script (a scripting language for the WAP browser). WAP browsers access WML and WML Script content by specifying a URL. Furthermore, WAP browsers can access dynamic content via CGI scripts and Java Servlets on the server.

While WML is suitable for rendering on a variety of browsers, it is best suited for browsers that have a limited display capability. WAP browsers tend to be used on small, hand-held hardware platforms such as cellular phones and beepers. Due to the difference in screen size between the typical HTTP browser and the typical WAP browser, there is a fundamental difference between HTML and WML. Whereas an HTML resource defines a virtual page, a WML resource defines a deck. A *deck* is a collection of cards, and a *card* maps to a virtual screen's worth of content on a WAP browser. If the amount of data represented by a card is greater than the physical size of a WAP browser's screen, the end user can scroll the display in order to see extra data.

WAP Software Architecture

Life might have been easier if the wireless Internet had been capable of using the same protocols as the wired Internet. Unfortunately, the characteristics that make TCP such a great wired protocol make it a weak protocol for wireless communications. The data integrity and the windowing features of TCP make it chatty, and chattiness is undesirable in a wireless protocol. TCP is a vastly successful protocol, however, and standard solutions for open systems are critically important. As a result of this paradox, many wireless standards evolved around IP, yet many proprietary solutions did not.

NOTE

Wireless networking is expensive. Either you pay a low monthly rate to get relatively little bandwidth, or you pay a per-minute charge for ample bandwidth. Either way, when you are using a wireless connection, you want to get your work done fast without a great deal of chattiness. While TCP has many features that make it a great general-purpose transport, TCP is extremely chatty.

Wireless devices that are built upon IP standards will tend to not perform as well as devices that are more bandwidth friendly. These devices are ready to participate on the Internet, however, provided that the bearer has Internet connectivity. Devices that are not built upon IP standards require a gateway in order to exchange packets with computers on the Internet. This problem is solved in part by the WAP gateway discussed previously.

In order to surf the Internet from a wireless device, you would want to have a browser that works consistently across a wide range of devices. A common browser implies a common protocol stack, however. Because some wireless devices are IP based and a great many have proprietary stacks, a common stack had to be created. Reliability and security features already built into Internet networking had to be reinvented in a package that had a reasonably small footprint, was capable of serving a broad spectrum of competing wireless datalinks, and performed well in a wireless environment. These requirements lead to the WAP architecture. Figure 2.11 shows both the IP and the non-IP mapping of the WAP architecture onto the ISO/OSI model.

The WAP architecture consists of six layers that we will briefly discuss next.

Application	WAE	WAE
Presentation		
Session	WSP	WSP
Transport	WTP	WTP
	UDP	WTLS
		WDP
Network	IP	Bearer Network
Data Link	Bearer Network	
Hardware		

Figure 2.11 WAP software architecture.

Bearer Network

As discussed previously, wireless devices exist with a variety of IP and non-IP based infrastructures. The primary function of the bearer network from a WAP perspective is to connect individual wireless devices to a WAP gateway. By performing this service, the WAP gateway can move traffic between the bearer network and the Internet, thereby providing Internet traffic to the wireless devices. The upper layers of the WAP protocol stack do not expect much from the bearer network in terms of security and/or reliability.

Wireless Datagram Protocol

The *Wireless Datagram Protocol* (WDP) [WAP WDP] provides essentially the same services as the UDP in an IP network. In fact, the services are so similar that an IP-based bearer network will use UDP in place of WDP. Recall from our earlier discussion that UDP provides a logical network-delivery scheme that enables one device to communicate with another.

Wireless Transport Layer Security

The *Wireless Transport Layer Security* (WTLS) [WAP WTLS] is the layer in the WAP architecture that is responsible for data integrity, privacy, and authentication. WTLS is similar to the industry-standard *Transport Layer Security* (TLS) [RFC 2246] security protocol; however, it has been optimized for use on bandwidth-constrained networks.

Wireless Transport Protocol

The *Wireless Transport Protocol* (WTP) [WAP WTPS] is a lightweight transaction-oriented transport protocol. This layer provides three classes of communications: unreliable one-way messaging, reliable one-way messaging, and reliable bidirectional transactions. As with the WTLS, the WTP has been optimized for use in bandwidth-constrained environments.

Wireless Session Protocol

The *Wireless Session Protocol* (WSP) [WAP WSPS] provides its upper layers with two forms of services: connection-oriented service and datagram service. Today, the WSP consists of a service that is particularly suited for communicating with an HTTP server: *WSP for Browsers* (WSP/B). Besides being optimized for use on wireless networks, the WSP/B contains special features for suspending and resuming connections.

NOTE

From our discussion of HTTP, recall that the header of an HTTP request or response is represented in ASCII. You can think of WSP/B as a binary version of HTTP. By tokenizing common contents of the HTTP/1.1 header with binary encoding bandwidth, we save bandwidth.

Wireless Application Environment

The *Wireless Application Environment* (WAE) is the layer of the WAP architecture that is of greatest interest to the WAP Servlet developer. At this layer, we find the applications that use the WSP to communicate with HTTP servers on the Internet. Here, we find applications that pertain to specific types of WAP clients (phones, pagers, and so on); however, we find one application of prime interest with regard to this book: WAP Browser.

Conclusion

In this chapter, we discussed a little about the hardware and software that make up the WAP architecture. This information, coupled with a bit of historical perspective, helped to establish the fundamental knowledge that is useful to the WAP Servlet developer. With these preliminary topics behind us, in the next chapter we will begin to develop code by examining WML.

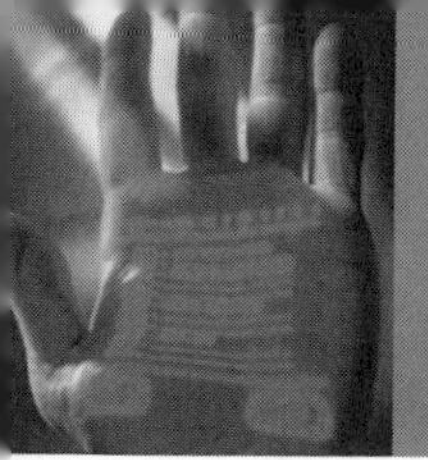

CHAPTER 3

Wireless Markup Language (WML)

Wireless Markup Language (WML) is the basic content-delivery mechanism for WAP. This chapter describes the features of WML in detail.

eXtensible Markup Language (XML)

Before getting into the details of WML, you should become familiar with the basics of *eXtensible Markup Language (*XML). While XML is a markup language in its own right, it also provides an excellent foundation for defining other markup languages.

Elements

Well-formed XML documents contain one element known as the root. The root element itself, however, can contain other elements—because elements can be embedded. Elements begin with a start tag and conclude with an end tag. Start tags are defined by the name of the element enclosed in angle brackets (< and >). For example, a start tag for an element with the name *TITLE* would be written as follows: `<TITLE>`. An end tag is distinguished from a start tag by a forward slash (/) preceding its name. For example, the tag that closes the element name *AUTHOR* would be written as follows: `</AUTHOR>`. The name of an element as it appears within its start and end tags is case sensitive.

Elements have content. As discussed previously, elements can be embedded; therefore, elements can have other elements as content. Additionally, elements can also contain text content. Regardless of the content (text and/or elements), all content of an element falls between its start and end tags. Although white space is significant in XML, it is not significant in the WML derivative. For readability, white space (spaces, tabs, carriage returns, and/or blank lines) can be freely embedded within the content of an element. Figure 3.1 shows an XML document that describes a book.

Referring to Figure 3.1, you will note that the BOOK element contains five subelements: *TITLE*, *AUTHORS*, *DESCRIPTION*, *ISBN*, and *REVIEWS*. In turn, the *AUTHORS* element contains two *AUTHOR* elements. The following elements contain text content: *TITLE*, *AUTHOR*, *DESCRIPTION*, and *ISBN*.

When an element is empty, a simple tag that replaces both the start and end tags can represent the element. This form of a tag is similar to an end tag, except that the forward slash follows the name. As shown in Figure 3.1, the empty element called *REVIEWS* is written as `<REVIEWS/>`.

Attributes

Besides content, elements can have attributes. Attributes provide a means of conveying additional information about an element. Attributes have a name-value pair (enclosed in double quotes) delimited by an equals sign. For example, an attribute called `currency` with a value of `us-dollars` would be written as follows: `currency="us-dollars."` An element can have no, one, or many attributes, and the attributes are defined within the start tag of the element that they augment. Figure 3.2 shows an XML document that contains elements with attributes.

Specifying the WML DTD

As stated previously, WML is a derivative of XML. In order for parties to agree on the meaning of an XML derivative, they need to describe the elements and their attributes. This description is embodied in a *Document Template Definition* (DTD). In the case of

```
<BOOK>
  <TITLE>Title</TITLE>
  <AUTHORS>
    <AUTHOR>Author 1</AUTHOR>
    <AUTHOR>Author 2</AUTHOR>
  </AUTHORS>
  <DESCRIPTION>
    Description
  </DESCRIPTION>
  <ISBN>0-000-00000-0</ISBN>
  <REVIEWS/>
</BOOK>
```

Figure 3.1 A sample XML document.

```
<product>
  <description>Big Blue Ball</description>
  <price currency="us-dollars">1.25</price>
  <sale currency="us-dollars" ends="JAN 19, 2000">0.75</sale>
</product>
```

Figure 3.2 An XML sample with attributes.

```
<?xml version="1.0"?>
<!DOCTYPE wml PUBLIC "-//WAPFORUM//DTD WML 1.1//EN"
    "http://www.wapforum.org/DTD/wml_1.1.xml">
```

Figure 3.3 An XML DTD.

WML, the WAP forum has published a DTD. In particular, this book deals with version 1.1 of the WML DTD.

After you have mastered the basics of WML by completing this chapter, you might prefer to use Appendix A, which contains a reference manual for WML. The reference material includes the 1.2 specification. Currently, most WAP browsers are at version 1.1; however, by the time this text is published, version 1.2 might be more popular. There are actually few changes with respect to WML.

When you create a static WML document in a file or dynamically render a WML document from a Servlet, the item must begin with the statements shown in Figure 3.3. The first line specifies that the document is a version 1.0 XML document. The next two lines specify the DTD in use: WML version 1.1. The WAP gateway and/or browser uses the DTD information to determine how the remainder of the document will be interpreted.

WML

In the early days of the WAP forum, browser manufacturers expressed an interest in having their WML implementation resemble the look and feel of other client features. For this reason, the WAP forum placed relatively few constraints on WML implementation. Thus, while WML itself is relatively straightforward, the most complicated part of learning WML is understanding how it varies from client to client and browser to browser. The remainder of this chapter introduces the features of WML, and when appropriate, we discuss significant client/ browser implementation differences.

NOTE

The behavior of your WML documents can vary between clients, browsers, and gateways. You should always test your implementation against more than one WAP emulator. Attempting to use your favorite emulator to the exclusion of all others will

result in an implementation that looks good on the targeted platform and will most likely be disappointing on other platforms.

Decks and Cards

Authors of WML documents are essentially constructing a deck of cards, where a card represents a virtual page of information from the perspective of the WAP browser and a deck is a collection of one or more cards that are closely related. The following sections discuss the basic elements used to construct decks and cards by using WML.

The <wml> Element

As discussed previously, WML documents are actually XML documents constructed according to the WML DTD. According to the DTD, the root element of a WML document is `<wml>`. The `<wml>` element must include one or more `<card>` elements (refer to the section concerning the `<card>` element). Figure 3.4a shows how the `<wml>` element is used to construct a simple WML document that is composed of a single card containing a paragraph of display text. Figure 3.4b shows a screen shot of the output generated by this example.

The <card> Element

The `<card>` element is used to define a virtual page of information to be displayed on a WAP browser. A card is a virtual page in the sense that its content can exceed the number of columns and/or rows available on the physical display. Refer to Figures 3.4a

```
<wml>
  <card>
    <p>Hello, world.</p>
  </card>
</wml>
```

Figure 3.4a The requisite "hello world" application.

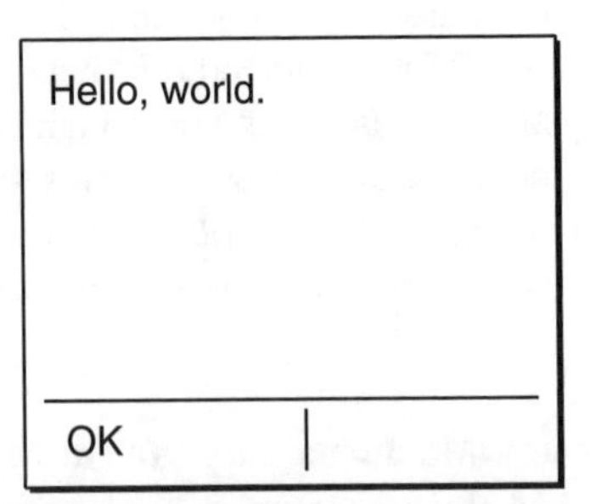

Figure 3.4b The obligatory screen shot of the requisite "hello world" application.

and 3.4b for an example of using the <card> element without attributes. In addition to this simple format, the behavior of the <card> element can be augmented through the use of seven attributes: id, title, newcontext, ordered, onenterforward, onenterbackward, and ontimer. The following sections demonstrate how you can use these attributes.

The id Attribute

The id attribute is used to specify an internal anchor point relative to the URL that identifies the deck. In general, Figure 3.5 shows how the id attribute is used to establish an anchor point on a card. In particular, this example shows an anchor point called card1 being established.

The title Attribute

The title attribute is used to specify the title of a card. The meaning of the title attribute is specific to the browser implementation. For example, the Phone.com browser uses the title attribute as the default name when generating a bookmark on a card. The Nokia browser uses the title attribute to place a title on the top line of the display. In general, Figure 3.6 shows how the title attribute is used to establish a title on a card. In particular, this example shows a title of *hello* being established.

The newcontext Attribute

The newcontext attribute is used to clear the browser history and reset variable values. The newcontext attribute takes one of two values: *true* or *false*. The implied

```
<wml>
  <card id="card1">
    <p>Hello</p>
  </card>
</wml>
```

Figure 3.5 How to use the id attribute on the <card> element.

```
<wml>
  <card title="hello">
    <p>Hello</p>
  </card>
</wml>
```

Figure 3.6 How to use the title attribute on the <card> element.

```
<wml>
  <card newcontext="true">
    <p>Hello</p>
  </card>
</wml>
```

Figure 3.7 How to use the `newcontext` attribute on the `<card>` element.

```
<wml>
  <card ordered="false">
    <p>Hello</p>
  </card>
</wml>
```

Figure 3.8 How to use the `ordered` attribute on the `<card>` element.

value of the `newcontext` attribute is *false*. Figure 3.7 shows how the `newcontext` attribute is used to clear the browser's history stack.

The ordered Attribute

The `ordered` attribute is used to specify whether or not the content of a card is ordered. The `ordered` attribute takes one of two values: *true* or *false*. The default value is *true*. Figure 3.8 shows how the `ordered` attribute is used to establish a card's content as unordered. We discuss the `ordered` attribute in greater detail next.

The onenterforward Attribute

The `onenterforward` attribute is used to specify an alternate URL to be loaded when the card at hand is navigated to by way of a `<go>` element (refer to the `<go>` element section as follows). The value of the attribute is the desired URL. In general, the WML document in Figure 3.9a shows how the `onenterforward` attribute is used to load an alternate card. Specifically, the example consists of a deck containing two cards: card1 and card2. An `onenterforward` attribute on card1 forwards control to card2. Figure 3.9b shows a progression of screen shots associated with the example. When navigating to card1, control is passed to card2 via the `onenterforward` attribute. Once on card2, pressing the back key will pass control to the previous card on the history stack (card1).

NOTE

Some WAP browsers do not provide a back button by default. The `<template>` element is used here and in other examples to provide this functionality. For now, you can ignore the `<template>` element and its content.

The onenterbackward Attribute

The `onenterbackward` attribute is used to specify an alternate URL to be loaded when the card at hand is backed into via the history stack. The value of the attribute is

```
<wml>
  <template>
    <do type="prev" name="prev">
      <prev/>
    </do>
  </template>
  <card id="card1" onenterforward="#card2">
    <p>This is card 1</p>
  </card>
  <card id="card2">
    <p>This is card 2</p>
  </card>
</wml>
```

Figure 3.9a Example of the `onenterforward` attribute of the `<card>` element.

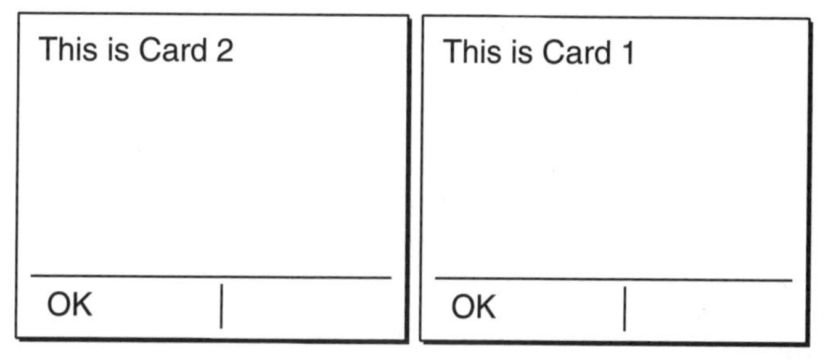

Figure 3.9b A screen-shot progression for the `onenterforward` attribute.

the desired URL. In general, the WML document in Figure 3.10a shows how the `onenterbackward` attribute is used to load an alternate card. Specifically, the example shows control being passed to card3 when backing into card1 from card2. Figure 3.10b shows a progression of browser screens associated with the example. Control passes from card1 to card2 by following the link to card2 (scrolling to the card2 link and pressing the *accept* key). When the *back* key is pressed to navigate back to card1, however, the `onenterbackward` attribute causes control to be passed to card3.

The ontimer Attribute

The `ontimer` attribute is used to specify an alternate URL to be loaded when the card at hand receives a timer event (refer to the `<timer>` tag). The value of the attribute is the desired URL. In general, the WML document in Figure 3.11a shows how the `ontimer` attribute is used to load an alternate card. Specifically, the example shows a deck containing two cards: card1 and card2. Control is passed to card2 when the timer on card1 expires. Figure 3.11b shows a progression of screen shots associated with the example. When navigating to card1, a timer is established for 50 tenths of a second. When the timer expires, the `ontimer` attribute causes control to be passed to card2.

```
<wml>
  <template>
    <do type="prev" name="prev">
      <prev/>
    </do>
  </template>
  <card id="card1" onenterbackward="#card3">
    <p>This is card 1</p>
    <p>You can go to <a href="#card2">card2</a>,
      but you will never get back.</p>
  </card>
  <card id="card2">
    <p>This is card 2</p>
  </card>
  <card id="card3">
    <p>This is card 3</p>
  </card>
</wml>
```

Figure 3.10a Example of the `onenterbackward` attribute of the `<card>` element.

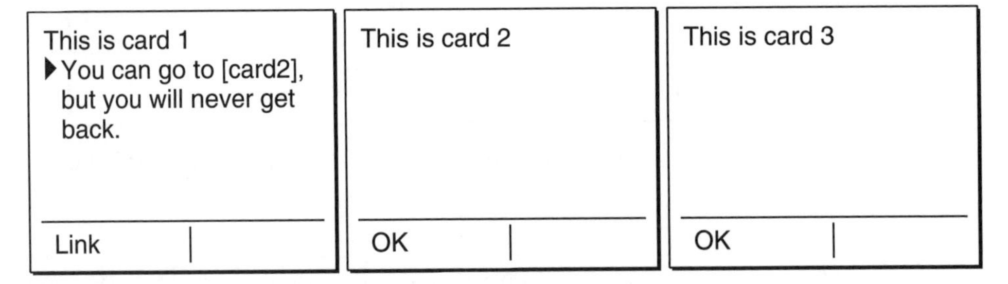

Figure 3.10b A screen-shot progression for the `onenterbackward` attribute.

Text Layout

As with HTML, the primary benefit of WML is publishing text. This section describes elements that are useful for presenting content.

The <p> Element

The <p> (paragraph) element is the fundamental WML text layout element. In fact, all <card> elements must contain at least one <p> element. Any content within the <p> element is displayed on the browser's display. The behavior of the <p> element can be modified via two attributes: `mode` and/or `align`. In general, the WML document in Fig-

```
<wml>
  <card id="card1" ontimer="#card2">
    <timer name="time" value="50"/>
    <p>Timer set.</p>
  </card>
  <card id="card2">
    <p>Timer expired!</p>
  </card>
</wml>
```

Figure 3.11a Example of the `ontimer` attribute for the `<card>` element.

Figure 3.11b A screen-shot progression for `ontimer`.

ure 3.12a shows how the `<p>` element is used to display text on the browser's display. Specifically, the example shows the text Hello being displayed. Figure 3.12b shows the browser screen associated with the example.

The mode Attribute

The `mode` attribute is used to specify the mode of a paragraph. The `mode` attribute can be set to one of two values: *wrap* or *nowrap*. The implied value of the `mode` attribute is *wrap*. If the *wrap* form of the `mode` attribute is used, text will wrap as it approaches the right margin of the browser's display. If the *nowrap* form of the `mode` attribute is used, text will marquee on a single line rather than wrap to subsequent lines. When wrapping text, a WAP browser will attempt to break lines between words. In general, the WML document in Figure 3.13a shows how the `mode` attribute is used. In particular, the example shows two paragraphs: one wrapped and one unwrapped. Figure 3.13b shows the browser screen associated with the example.

TIP

When displaying several short lines of text that are formatted similarly on consecutive lines, you should use *nowrap* mode. When a large block of text is displayed, you should use *wrap* mode. Reading lengthy output in marquee format is difficult on most WAP browsers.

```
<wml>
  <card>
    <p>Hello</p>
  </card>
</wml>
```

Figure 3.12a Example of the basic <p> element.

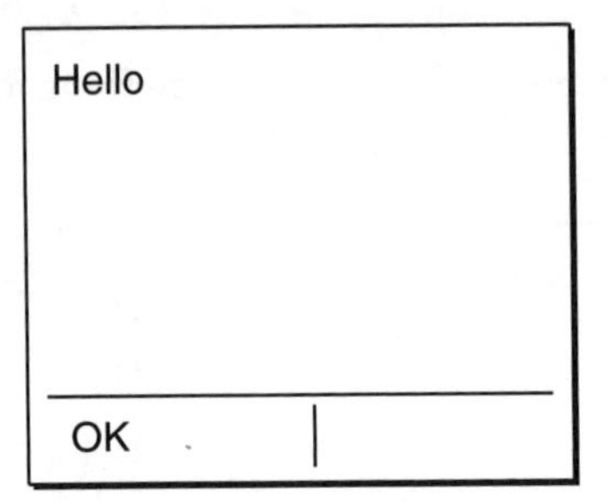

Figure 3.12b A screen shot of the basic <p> element.

```
<wml>
  <card>
    <p mode="nowrap">Line 1 is long enough to wrap but wont.</p>
    <p mode="wrap">Line 2 is long enough to wrap and it will.</p>
  </card>
</wml>
```

Figure 3.13a Example of the `mode` attribute for the <p> element.

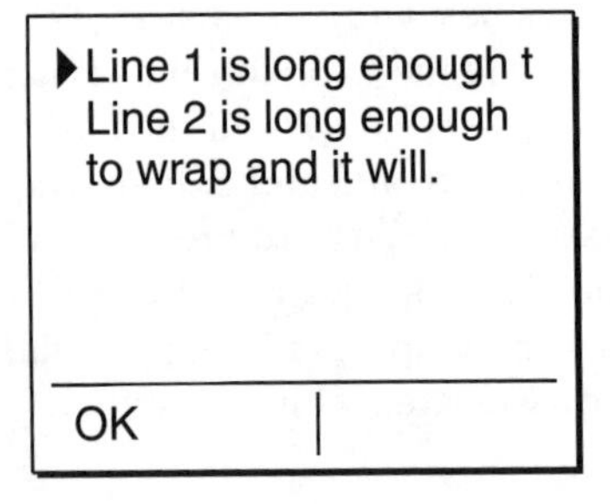

Figure 3.13b Screen shots of the `mode` attribute example.

The align Attribute

The `align` attribute is used to specify the alignment of text within a paragraph. The `align` attribute can be set to one of three values: *left*, *right*, or *center*. The implied

```
<wml>
  <card>
    <p align="left">left</p>
    <p align="center">center</p>
    <p align="right">right</p>
  </card>
</wml>
```

Figure 3.14a Example of the `align` attribute for the `<p>` element.

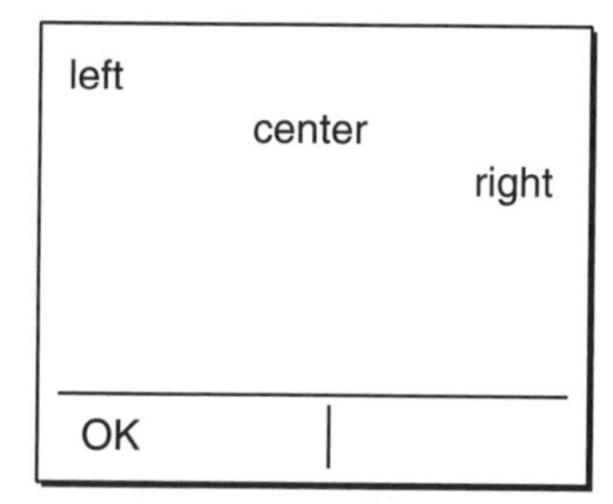

Figure 3.14b Screen shot of the `align` attribute example for the `<br/>` element.

value of the `align` attribute is *left*. In general, the WML document in Figure 3.14a shows how you can use the `align` attribute. Specifically, the example shows three paragraphs that use different alignments. Figure 3.14b shows the browser screen associated with the example.

The
 Element

The `<br/>` (break) element is used to force a line break within a paragraph. The `<br/>` element has neither content nor attributes. In general, the code segment in Figure 3.15a shows how you can use the `<br/>` element. Specifically, the example shows two paragraphs that use the `<br/>` element to insert line breaks. Figure 3.15b shows the browser screen associated with the example.

Rather than using the `<br/>` element to insert breaks in paragraph text, you can use multiple `<p>` elements—because each paragraph starts on its own line. The `<br>` element used within a paragraph, however, preserves the paragraph's mode and alignment.

The Formatting Elements

Several tags are useful for changing the appearance of text within a paragraph. While the actual behavior of these tags is browser dependent, Table 3.1 describes the objective of the formatting elements:

```
<wml>
  <card>
    <p mode="nowrap">Last Name:<br/>Cook</p>
    <p mode="nowrap">First Name:<br/>John</p>
  </card>
</wml>
```

Figure 3.15a Example of the mode attribute for the <p> element.

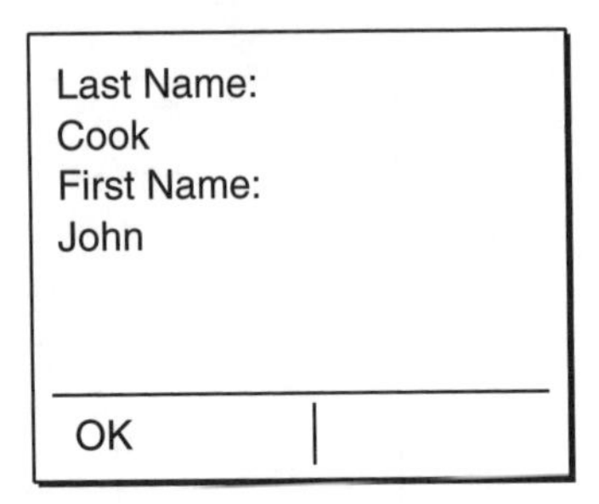

Figure 3.15b Screen shot of the mode attribute example.

Table 3.1 Text Formatting Elements

ELEMENT	FUNCTION	COMMENT
<em>	Emphasis	Uses some technique to emphasis text
<strong>	Strong	Displays text stronger than normal
<b>	Bold	Displays text in the bold form of the font
<i>	Italic	Displays text in the italic form of the font
<u>	Underscore	Displays text in the underlined from of the font
<big>	Big	Displays text in a larger than normal pitch
<small>	Small	Displays text in a smaller than normal pitch

You can use formatting elements in WML decks wherever text is displayed. In general, the WML document in Figure 3.16a shows how the formatting elements are used. Specifically, the example shows a paragraph with each of the names of the seven formatting elements formatted accordingly. Figure 3.16b shows the browser screen associated with the example.

NOTE

The exact meaning of the format elements varies between clients. You should become familiar with how the format elements behave on the clients that you intend to target.

```
<wml>
  <card>
    <p>normal</p>
    <p>
      <em>em </em>
      <strong>strong </strong>
      <b>b </b>
      <i>i </i>
      <u>u </u>
      <big>big </big>
      <small>small </small>
      <b><i>b&amp;i</i></b>
    </p>
  </card>
</wml>
```

Figure 3.16a Example of the format elements.

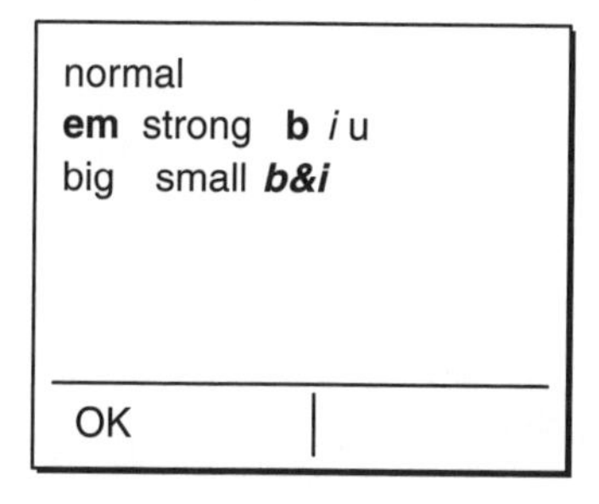

Figure 3.16b Screen shot of the format elements example.

Event Bindings

Two WML elements fall under the general category of event bindings: <do> and <onevent>. This section examines both of these elements by exploring their attributes and providing examples.

The <do> Element

The <do> element is used to associate actions with browser functions. These functions are associated with hard keys, soft keys, or system menu items. On most WAP clients, the display is situated above the soft keys. The bottom line of the display is used to display the label associated with a soft key directly above the key. The content of the <do> element must be one of the following elements: <go>, <prev>, <noop>, or <refresh>. The function of each of these tags is discussed in respective sections.

NOTE

Examples in this section use the <go> element (refer to the <go> element on page 49).

The type Attribute

The <do> element has one required attribute: type. The value of the type attribute is typically one of the following: *accept, options, delete, help, prev,* or *reset.* If the value is *accept,* the <do> element is associated with the primary soft key (the *accept* key). In general, the WML document shown in Figure 3.17a shows an example of this situation. In particular, the example deck contains two cards: card1 and card2. A <do> element is used on card1, associating card2 with the *accept* key. Figure 3.17b shows a progression of screen shots associated with the example.

If the value of the type attribute is options, the <do> tag is associated with the secondary soft key (the *options* key). In general, the code segment in Figure 3.18a shows an example of this situation. In particular, the example deck contains two cards: card1 and card2. A <do> element is used in card1, associating card2 with the *options* key. Figure 3.18b shows a progression of screen shots produced by the example. The first panel shows card1. On card1, the *options* key is associated with card2's URL via the <do> element. Once the *options* key is pressed, card2 is displayed as shown in panel two.

```
<wml>
  <card id="card1">
    <do type="accept">
      <go href="#card2"/>
    </do>
    <p>This is card 1</p>
  </card>
  <card id="card2">
    <p>This is card 2</p>
  </card>
</wml>
```

Figure 3.17a Example of the type attribute for the <do> element.

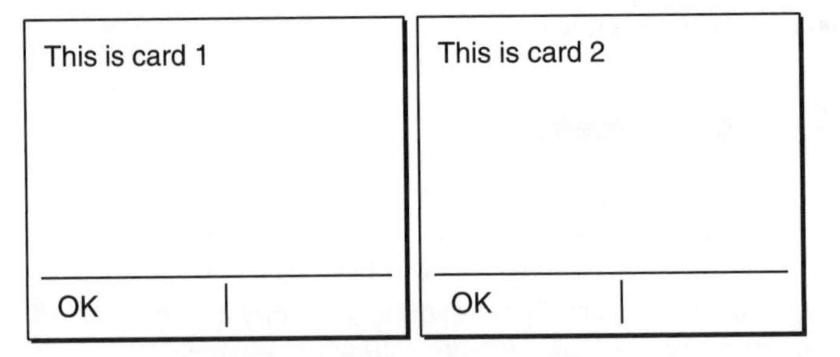

Figure 3.17b Screen shot progression for the type example.

```
<wml>
  <card id="card1">
    <do type="options">
      <go href="#card2"/>
    </do>
    <p>This is card 1</p>
  </card>
  <card id="card2">
    <p>This is card 2</p>
  </card>
</wml>
```

Figure 3.18a Example of the `type` options attribute of the `<do>` element.

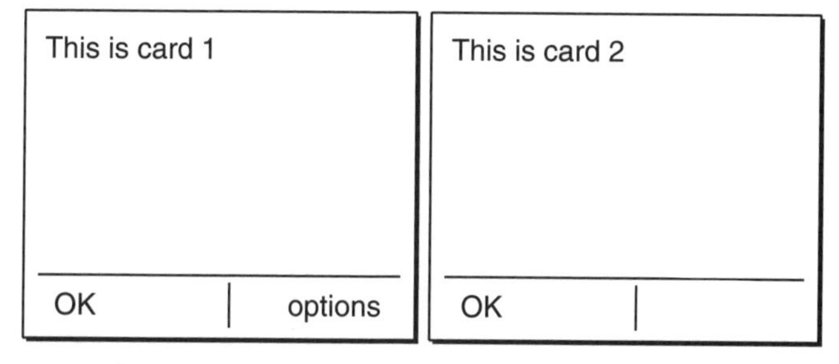

Figure 3.18b Screen shot progression for the `type` options example.

NOTE

While all WAP browsers must implement one soft function key, there is no requirement for a second soft function key. Most browsers have two or more soft function keys. On some browsers, the *options* key is implemented on the primary soft key. If two or more `<do>` elements (other than type *prev*) are defined, they are placed in a dynamic menu associated with the primary key (labeled options).

TIP

Because some browsers implement the options key as a menu pick under a soft key called label, you should allow the browser to label the *option* type `<do>` element. If you feel compelled to label the `<do>`, you should avoid labeling option, because that action might result in the user having to select something labeled *option* twice (which is somewhat confusing).

If the value of the `type` attribute is *delete*, the `<do>` tag is associated with the *delete* key. In general, the code segment in Figure 3.19a shows an example of this situation. In particular, the example deck contains two cards: card1 and card2. A `<do>` element is used

```
<wml>
  <card id="card1">
    <do type="delete">
      <go href="#card2"/>
    </do>
    <p>This is card 1</p>
  </card>
  <card id="card2">
    <p>This is card 2</p>
  </card>
</wml>
```

Figure 3.19a Example of the `type delete` attribute of the `<do>` element.

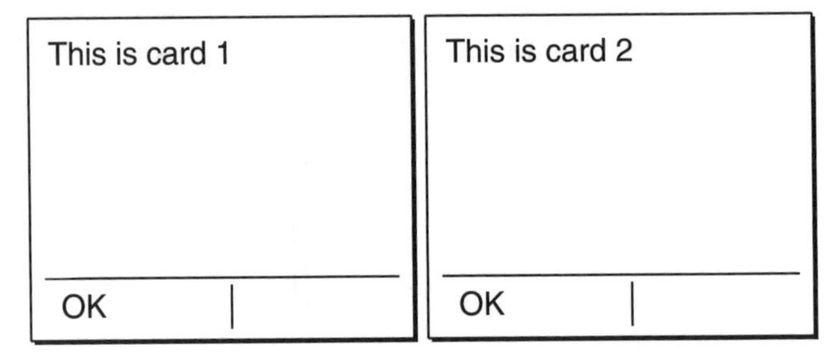

Figure 3.19b Screen shot progression of the `type delete` example.

in card1, associating card2 with the *delete* key. Figure 3.19b shows a progression of screen shots associated with the example. The first screen shows card1. Upon pressing the *delete* key, the second screen will be shown due to the <do> element on card1 that associates the *delete* key with card2.

NOTE

The delete function is often tied to the hard key that is used to clear a character. The clear hard key is labeled with one of the following labels: C, CLR, or CLEAR. On some browsers, the clear key doubles as the *back* key. If the delete function is not available due to a key conflict, the delete function should either be implemented as a soft key labeled "delete" or as an entry labeled "delete" in a dynamic menu implemented as a soft key. Many browsers do not implement the delete function; however, if there is a *clear* key conflict (hence the following tip).

TIP

Developers should always provide their own delete function implemented as a soft key or as a menu entry associated with a soft key. Because most developers do not use the *delete* type <do> element, many users will not know to look for the *delete*

key in order to perform a delete function. Furthermore, many browsers do not provide a way of executing a delete function; therefore, failure to explicitly implement a delete function might lead to confusion and/or an inability to delete. Although a delete function should be explicitly implemented, users will appreciate the convenience and consistency that the *delete* type <do> element offers.

If the value of the type attribute is *help*, the <do> tag is associated with the *help* key. In general, the code segment in Figure 3.20a shows an example of this situation. In particular, the example deck contains two cards: card1 and card2. A <do> element is used in card1, associating card2 with the *help* key. Figure 3.20b shows a progression of screen shots associated with the example. Initially, card1 is shown. Upon pressing the *help* key, card2 will be shown due to the <do> element on card1 that associates the *help* key with card2.

NOTE

Many browsers do not have a *help* key. On these browsers, the functional equivalent of the *help* key is accessed via the *help* entry within the browser menu. The browser menu is usually accessed via a client key. If the browser does not implement a

```
<wml>
  <card id="card1">
    <do type="help">
      <go href="#card2"/>
    </do>
    <p>This is card 1</p>
  </card>
  <card id="card2">
    <p>This is card 2</p>
  </card>
</wml>
```

Figure 3.20a Example of the type help attribute of the <do> element.

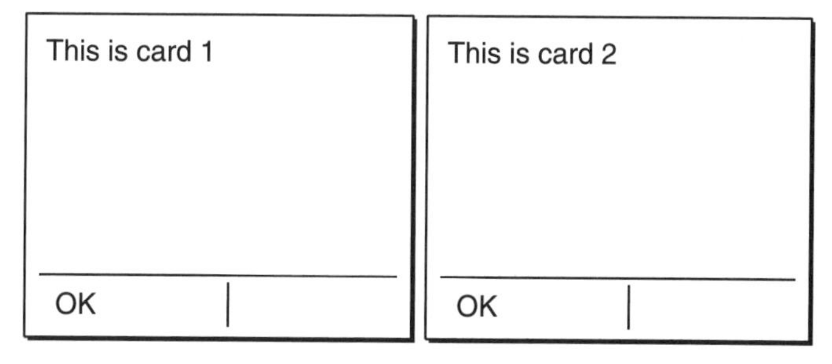

Figure 3.20b Screen shot progression of the type help example.

browser menu, the help function will either be associated with a soft key or a menu pick on a dynamic menu associated with a soft key.

If the value of the `type` attribute is *reset*, the <do> tag is associated with the reset function. Because most browsers do not have a hard key for the reset function, this function is typically implemented as a soft key (or a menu pick off of a soft key). In general, the code segment in Figure 3.21a shows an example of this situation. In particular, the example deck contains two cards: card1 and card2. A <do> element is used in card1, associating card2 with the *reset* key. Figure 3.21b shows a progression of screen shots that are generated by the example. The first screen shows card1 demonstrating that the reset function is implemented on a soft key. Upon pressing the *reset* key, you will display card2 due to the <do> element on card1 that associates the *reset* key with card2.

So, what would happen if no soft key were available for the reset function? For example, if there were two <do> elements (one each for the `type` attribute values *options* and *reset*) on a browser with two soft keys, how would the conflict for the options soft key be resolved? In this event, the browser will create a menu function and associate it with the options soft key. Both the options and the reset functions would become menu picks on that menu. In general, the code segment in Figure 3.22a shows an example of this situation. In particular, the example deck contains two cards: card1 and card2. Two

```
<wml>
  <card id="card1">
    <do type="reset">
      <go href="#card2"/>
    </do>
    <p>This is card 1</p>
  </card>
  <card id="card2">
    <p>This is card 2</p>
  </card>
</wml>
```

Figure 3.21a Example of the `reset type` attribute for the <do> element.

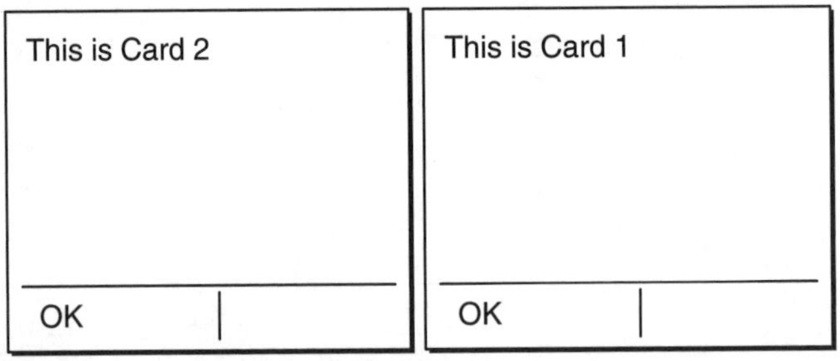

This is Card 2	This is Card 1
OK	OK

Figure 3.21b Screen shot progression of the `reset type` example.

```
<wml>
  <template>
    <do type="prev" name="prev">
      <prev/>
    </do>
  </template>
  <card id="card1">
    <do type="reset">
      <go href="#card2"/>
    </do>
    <do type="options">
      <go href="#card2"/>
    </do>
    <p>This is card 1</p>
  </card>
  <card id="card2">
    <p>This is card 2</p>
  </card>
</wml>
```

Figure 3.22a Example of option key conflict resolution.

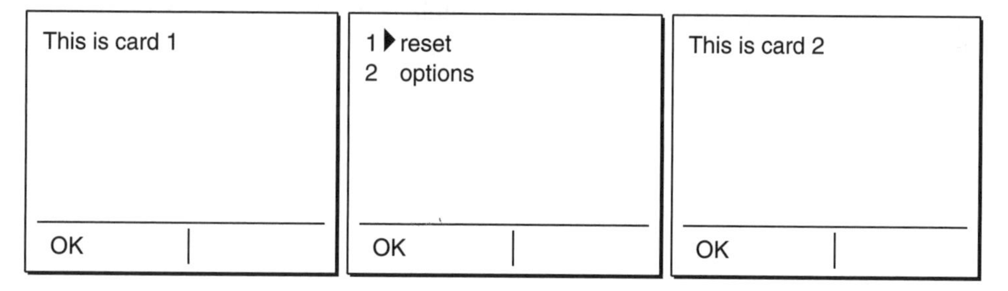

Figure 3.22b Screen shot progression for the conflict resolution example.

<do> elements are used in card1, associating card2 with the *reset* and *options* keys, respectively. Figure 3.22b shows a progression of screen shots associated with the example. Initially, card1 is displayed, demonstrating that a menu has been implemented on a soft key. Upon pressing the *menu* key, you will display the menu that resolves the conflict. Upon selecting either entry in the menu, you pass control to card2.

The label Attribute

While the WAP browser does a reasonable job of labeling soft keys and functions, developers frequently need to label their <do> element in order to clarify the meaning of the function to the end user. The `label` attribute takes a string value (the desired label). If the <do> element is associated with a soft key, the value of the `label` attribute becomes the soft key label (in other words, the name that appears above the soft key).

```
<wml>
  <template>
    <do type="prev" name="prev">
      <prev/>
    </do>
  </template>
  <card id="card1">
    <do type="accept" label="yes">
      <go href="#card2"/>
    </do>
    <do type="options" label="cancel">
      <go href="#card3"/>
    </do>
    <p>Are you sure you want to delete this record?</p>
  </card>
  <card id="card2">
    <p>You said yes.</p>
  </card>
  <card id="card3">
    <p>You canceled.</p>
  </card>
</wml>
```

Figure 3.23a Example of the `label` attribute for the `<do>` element.

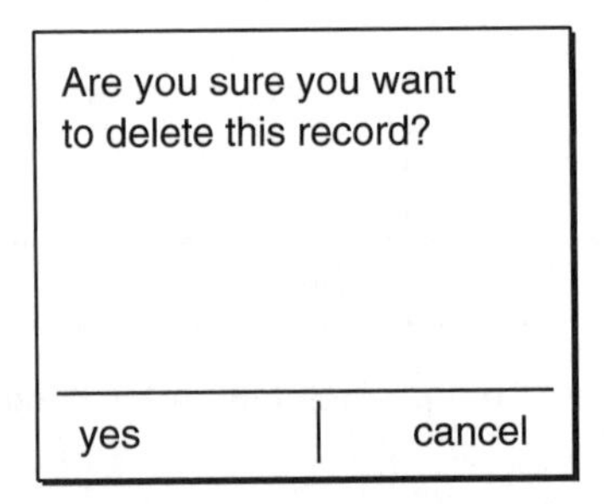

Figure 3.23b Screen shot progression for the `label` example.

If the <do> element is associated with a function that is implemented as a menu item, the value of the `label` attribute becomes the name of that menu item pick.

In general, the WML document in Figure 3.23a shows an example that demonstrates the use of the `label` attribute. In particular, the example deck contains three cards: card1, card2, and card3. Two <do> elements are defined on card1. The first <do> element has a `label` attribute with a value of *yes*; it associates the *accept* key with card2. The second <do> element has a `label` attribute with a value of cancel; it associates the options key with card3. Figure 3.23b shows a screen shot of card1 from the example.

NOTE

On most WAP browsers, all soft key labels share a line on the display; therefore, names should be short. Choosing names that are six characters or fewer should enable a Servlet to look correct on most any browser.

The name Attribute

The name attribute is used in conjunction with <do> elements defined within a deck's <template> element. See the <template> element as follows for a description and example of the name attribute.

The optional Attribute

The optional attribute can take one of two values: *true* or *false*. The implied value is *false*. When the optional attribute has a value of *true*, a browser is permitted to ignore the <do> element.

The <onevent> Element

The <onevent> element is similar to the <do> element. In fact, the difference is subtle. Where the <do> element deals with cause (for example, pressing the back button), the <onevent> element deals with effect (for example, backing into a card). The <onevent> element is used to associate an action with an event. The content of the <onevent> element must be one of the following elements: <go>, <prev>, <noop>, or <refresh>. The function of each of these tags is discussed in respective sections as follows. As with the <do> element, the <go> element appears in the following examples.

The type Attribute

The <onevent> element has one required attribute: type. The value of the type attribute is one of the following: *onpick*, *onenterforward*, *onenterbackward*, or *ontimer*. If the value is *onpick*, the <onevent> element is associated with the selection of an item from a <select> element. Due to the complexity of the <select> element, we will defer the discussion of the type attribute with a value of onpick (refer to the <select> element as follows).

If the value of the type attribute is *onenterforward*, the <onevent> element will react to the browser's navigating forward into a card. Therefore, the behavior is similar to using the onenterforward attribute in the <card> element. The primary difference is that the content of the <onevent> provides greater flexibility. In general, the code segment in Figure 3.24 shows an example. In particular, the deck contains two cards: card1 and card2. An <onevent> element is used in card1 to navigate to card2.

If the value of the type attribute is *onenterbackward*, the <onevent> element will react to the browser's navigating backwards into a card. Thus, the behavior is similar to

```
<wml>
  <card id="card1">
    <onevent type="onenterforward">
      <go href="#card2"/>
    </onevent>
    <p>
      You won't see this card.
    </p>
  </card>
  <card id="card2">
    <p>
      You will see this one.
    </p>
  </card>
</wml>
```

Figure 3.24 Example of the type `onenterforward` attribute for the `<onevent>` element.

using the `onenterbackward` attribute in the `<card>` element. The primary difference is that the content of the `<onevent>` element provides greater flexibility. In general, the code segment in Figure 3.25 shows an example. In particular, the element deck contains three cards: card1, card2, and card3. A `<onevent>` element is used in card1 to navigate to card3 when backing out of card2.

If the value of the `type` attribute is *ontimer*, the `<onevent>` element will react to a timer's expiration. Therefore, the behavior is similar to using the `ontimer` attribute in the `<card>` element. The primary difference is that the content of `<onevent>` provides greater flexibility. In general, the code segment in Figure 3.26 shows an example. In particular, the element deck contains two cards: card1 and card2. A `<onevent>` element is used in card1 to navigate to card2 when the card's timer expires.

Deck Level Declarations

Besides the `<card>` element, two elements can be used within the content of the `<wml>` element: `<head>` and `<template>`. The following sections describe these elements and provide examples of their use.

The <head> Element

Although the `<head>` element is optional, if used, it must appear first in the content of the `<wml>` element. The sole purpose of the element is to contain `<access>` and `<meta>` elements. We describe these elements in the following sections.

```
<wml>
  <template>
   <do type="prev" name="prev">
     <prev/>
   </do>
  </template>
  <card id="card1">
    <onevent type="onenterbackward">
      <go href="#card3"/>
    </onevent>
    <p>This is card 1</p>
    <p>You can go to <a href="#card2">card2</a>,
      but you will never get back.
        </p>
  </card>
  <card id="card2">
    <p>This is card 2</p>
  </card>
  <card id="card3">
    <p>This is card 3</p>
  </card>
</wml>
```

Figure 3.25 Example of the type `onenterbackward` attribute for the `<onevent>` element.

```
<wml>
  <card id="card1">
    <onevent type="ontimer">
      <go href="#card2"/>
    </onevent>
    <timer name="time" value="50"/>
    <p>Timer set.</p>
  </card>
  <card id="card2">
    <p>Timer expired!</p>
  </card>
</wml>
```

Figure 3.26 Example of the type `ontimer` attribute for the `<onevent>` element.

The <access> Element

A *referent* is a card that refers to a deck or a card within a deck. The `<access>` element is used to limit deck access to particular referents. By default (when no `<access>` element is specified), access is granted to all referents. Access is limited by using one or both of the available attributes of the <access> element: domain and path. The value of the domain attribute is the DNS name of a system that can access the deck. The implied value of the domain attribute (in other words, when the attribute is not used) is the domain of the deck at hand. The value of the path attribute is the root file path of referents that can access the deck. The implied value of the path attribute (in other words, when the attribute is not used) is the path '/'. If the <access> element is used without either attribute, access to a deck is limited to other decks on the same server as the referred deck.

In general, Figure 3.27a demonstrates the use of the <access> element. Specifically, the example shows a deck. This deck has an <access> element that limits access to the "foo" domain. Assuming that your machine does not belong to this domain, attempting to access the deck will result in an access error. Figure 3.27b shows a screen shot depicting the typical access error that would be displayed upon attempting to access the deck in the example.

The <meta> Element

The <meta> element is used to send metadata to either the WAP client or gateway. Each <meta> element has a required property that is realized with one of two attributes: name or http-header. The form of the element that uses the name attribute is primarily intended to invoke a proprietary feature of a specific WAP client. In this case, the value of the content attribute specifies arguments to the named feature.

```
<wml>
  <head>
    <access domain="foo"/>
  </head>
  <card id="card1">
    <p>You can't get here from there!</p>
  </card>
</wml>
```

Figure 3.27a Example of the <access> element.

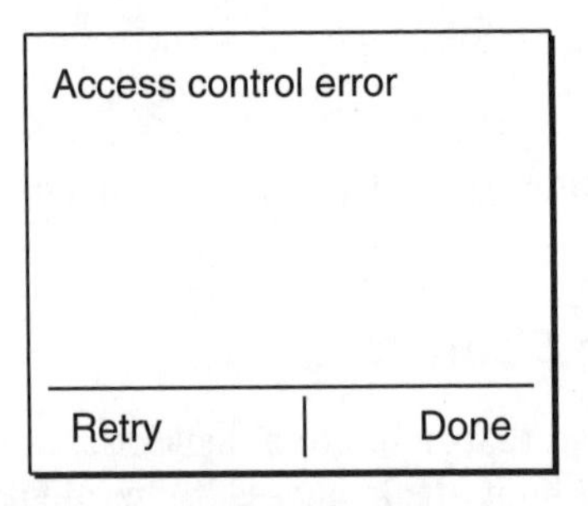

Figure 3.27b Screen shot of the <access> example.

The form of the element that uses the `http-header` is used to set a name-value pair within the HTTP header. In this case, the value of the `http-header` attribute is the name of the HTTP header item, and the value of the `content` attribute is the value. A logical question to ask here is why use this method to establish a HTTP name-value pair when (as we will see in Chapter 4), the Servlet can handle this task? Because the `<meta>` element is part of the WML payload, it can be evaluated by both the gateway and the WAP browser. An HTTP header can only be evaluated by the gateway. Secondly, if you are serving static WML decks from an HTTP server, you have no means at your disposal for specifying HTTP headers.

The most important use of the `<meta>` element is cache control. In this section, we will look at both the `expires` and `cache-control` headers. The `expires` header is used to specify a time and date after which the deck will no longer be valid. The following line shows an example of this header:

```
Expires: Thu, 29 Jun 2000, 10:35:00 GMT
```

The `cache-control` header can take a few forms. A value of `must-revalidate` specifies that a browser must revalidate the content when backing into a page.

NOTE

As with HTML, the default behavior for WML is that when backing up, you should use a cached version of a deck without validating the cache parameters. The `must-revalidate` forces validation of the cache parameters. If the parameters indicate that the page is still valid, the cache entry is used; otherwise, a new request is made for the content.

Another `cache-control` value is `no-cache`. The `no-cache` value specifies that a page should not be cached. In this case, while the deck will not go into cache, it will be placed on the history stack, so the `must-revalidate` must be used with `no-cache` if you want to always force the reloading of a particular deck in all cases. Yet another useful `cache-control` value is `max-age`. The `max-age` value is used to establish the number of seconds that the deck will be valid in the cache. The following lines show some examples of the `cache-control` element:

```
Cache-Control: must-revalidate
Cache-Control: max-age=3600
Cache-Control: must-revalidate, no-cache
Cache-Control: must-revalidate, max-age=3600
```

For information about other HTTP header name-value pairs that you can use for cache-control, refer to [RFC 2616]. Figure 3.28 shows an example of a WML deck that uses the `<meta>` method for cache control.

The <template> Element

Although the `<template>` element is optional, if you use it, the element must precede the first `<card>` element within the content of the `<xml>` element. When used, the `<template>` element defines the default behavior of cards within a deck. A simple

```
<wml>
  <head>
    <meta http-equiv="Expires"
    content="Thu, 29 Jun 2000, 10:35:00 GMT"/>
    <meta http-equiv="Cache-Control"
    content="must_revalidate, no-cache"/>
  </head>
  <card>
    <p>Hello, world.</p>
  </card>
</wml>
```

Figure 3.28 Example of the `<meta>` element.

form of behavioral modification can be achieved by specifying a URL for the value of the following `<template>` element attributes: `onenterforward`, `onenterbackward`, and/or `ontimer`. When using these attributes, the outcome is the same as if the same attribute were used on each card of the deck. (Refer to the description of `onenterforward`, `onenterbackward`, and `ontimer` under the description of the `<card>` element for examples of their usage.)

In addition to specifying attributes, you can also use the content of the `<template>` element to specify default behavior for all cards within the deck. You can use the following elements as content within the `<template>` element: `<do>` and `<onevent>`. (Refer to the `<do>` and `<onevent>` elements for examples of their use.)

If the `name` attribute is used to name a `<do>` element within the `<template>` element, the same name must be specified on a `<do>` statement at the card level in order to override the action association. If not, both the template and the card level associations will stand. In general, Figure 3.29a shows an example of this situation. Specifically, the example shows two cards: card1 and card2. A deck-level template associates both the *accept* and *options* key with card2. On card1, the association of the options key is overridden by name—replacing the deck-level association. Furthermore, on card1, an additional association is established on the *accept* key (not replacing the deck-level association because the name is not specified). Because a conflict exists for the *accept* key on card1, a dynamic menu is created for the redundant *accept* key association and the original options key association. Figure 3.29b shows a screen shot progression demonstrating the results.

Variables

WML variables offer a convenient way to save context information on the WAP client. They are a useful way to pass values between cards and decks. By using variables, developers are able to implement decks that behave as subroutines. Variables can be used as or in output text and/or attribute values. To access a variable, you must use the

```
<wml>
  <template>
    <do type="accept" label="t1" name="foo">
      <go href="#card2"/>
    </do>
    <do type="options" label="t2" name="bar">
      <go href="#card2"/>
    </do>
  </template>
  <card id="card1">
    <do type="accept" label="c1">
      <go href="#card2"/>
    </do>
    <do type="options" label="c2" name="bar">
      <go href="#card2"/>
    </do>
    <p>This is card1</p>
  </card>
  <card id="card2">
    <p>This is card2</p>
  </card>
</wml>
```

Figure 3.29a Example of the `<template>` element.

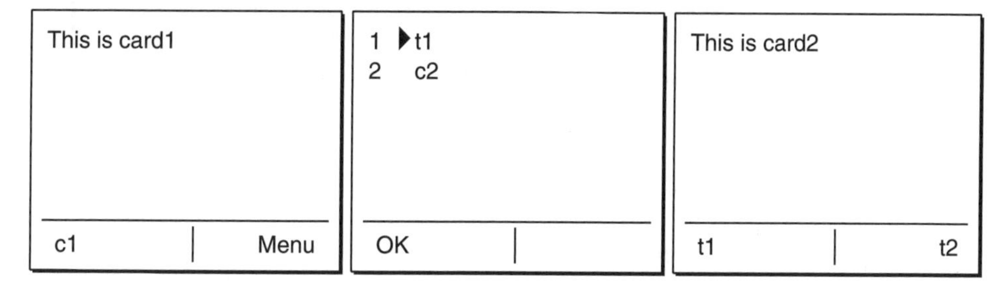

Figure 3.29b Screen shot of the `<template>` example.

following syntax: `$(var)`, where `var` is the name of a variable. To set the value of a variable, use the `<setvar>` element.

The <setvar> Element

The `<setvar>` element can be used within the content portion of the `<go>`, `<prev>`, and `<refresh>` elements to establish or reset the value of a variable. The element has two required attributes: `name` and `value`. The name of the variable is established via the `name` attribute. The value of the variable is established via the `value` attribute. In

```
<wml>
  <card id="card1">
    <do type="accept">
      <go href="#card2">
        <setvar name="var" value="is cool"/>
      </go>
    </do>
    <p>This is card 1</p>
  </card>
  <card id="card2">
    <p>This is card 2</p>
    <p>WAP $(var)</p>
  </card>
</wml>
```

Figure 3.30a Example of the <setvar> element.

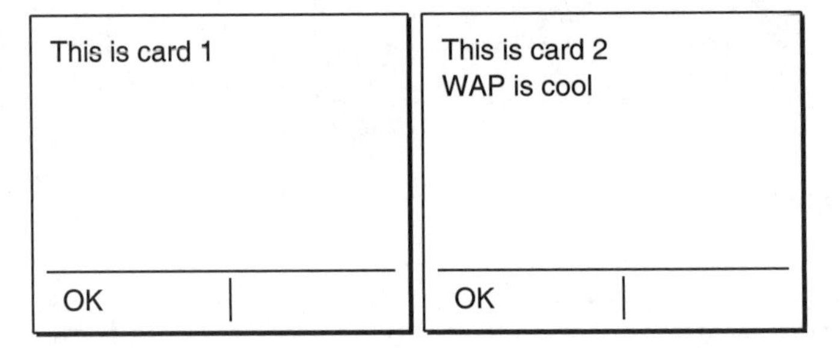

Figure 3.30b Screen shot of the <setvar> example.

general, Figure 3.30a shows how the element is used. Specifically, the example shows a deck of two cards: card1 and card2. On card1, a <do> element is used to associate the *accept* key with card2. Within the content of this <do> element, the <setvar> element is used to set the variable `$(var)` with the value "is cool." Figure 3.30b shows a screen shot progression associated with the example. The first panel shows the output generated by card1. When the *accept* key is pressed on card1, control is forwarded to card2. As shown in panel two, the output of card2 shows the value of the variable `$(var)` as established by the <setvar> element.

Tasks

Four elements fall under the general heading of tasks: <go>, <prev>, <refresh>, and <noop>. Readers will recall that these four elements are used as content within the <do> and <onevent> elements. The following sections demonstrate how these elements are used to tailor the behavior of the event-binding elements.

The <go> Element

The <go> element is used to move the browser to another deck or card. The <go> element can take any of the following attributes: href, sendreferer, and/or method. The href attribute is required. There is also an accept-charset attribute; however, this attribute is not required and is beyond the scope of this book.

The href Attribute

The href attribute provides a means of establishing a hypertext reference to a card or deck. The value of the href attribute is a valid absolute or relative URL. In general, Figure 3.31 shows an example of the href attribute. Specifically, the example shows that pressing the *accept* key will result in the page2 card of the current deck being displayed. Pressing the *options* key will result in the deck at relative URL options.wml being loaded. Pressing the *reset* key will result in the absolute URL www.wapform.org being loaded.

The sendreferer Attribute

The value of the sendreferer attribute can be one of two values: *true* or *false*. The default value is *false*. If the value is *true*, the referer header option in the resulting HTTP request will be set. The referer header option, if present, is the URL of the referent card. The referent card is the card that contained the <go> element. Having the referent information is useful for security and for accounting. From a security perspective, the Servlet that receives the HTTP request might choose to send HTTP responses only when the reference is a certain value. From an accounting prospective, Servlet

```
<wml>
  <card>
    <do type="accept">
      <go href="#page2"/>
    </do>
    <do type="option">
      <go href="options.wml"/>
    </do>
    <do type="reset">
      <go href="http://www.wapforum.org"/>
    </do>

    ...

  </card>
</wml>
```

Figure 3.31 Example of the href attribute of the <do> element.

```
<wml>
  <card id="card1">
    <do type="accept">
      <go
        href="http://localhost:8080/servlet/book.servlet.Dump.Dump"
         sendreferer="true"
         method="post">
        <postfield name="OVP" value="Yes"/>
      </go>
    </do>
    <p>Click ok to dump HTTP header.</p>
  </card>
</wml>
```

Figure 3.32a Example of the `sendreferer` attribute of the `<go>` element.

```
HTTP Header Options:
  Connection: keep-alive
  Date: Fri, 04 Feb 2000 15:27:45 GMT
  Content-Length: 7
  Content-Type: application/x-www-form-urlencoded
  Accept: text/vnd.wap.wml,text/vnd.wap.wmlscript,
    application/vnd.wap.wmlc, application/vnd.wap.wmlscriptc,
    image/vnd.wap.wbmp, image/gif
  Accept-Charset: UTF-8, ISO-8859-1, ISO-10646-UCS-2
  Host: 127.0.0.1:8080
  Referer: file:/C:/wap/test.wmlc
  User-Agent: Nokia-WAP-Toolkit/1.3beta
```

Figure 3.32b A screen shot progression for the `sendreferer` example.

authors might wish to track the number of HTTP requests that came from a particular referent card.

The method Attribute

The value of the `method` attribute must be either *get* or *post*. The implied value is get unless the `<go>` element contains a `<postfield>` element (refer to the `<postfield>` element). If the `<go>` element contains a `<postfield>` element, then the implied value of the `method` attribute is *post*. When the value of the `method` attribute is *get*, the `<go>` element appears as an HTTP get request. In a get request, the parameters of a request are appended to the query string of the URL. When the value of the `method` attribute is *post*, the `<go>` element appears as an HTTP post request. In a post request, the parameters of a request are sent as content in the HTTP request. In general, the WML document in Figure 3.33a shows an example of the `method` attribute. Specifi-

```
<wml>
  <template>
    <do type="prev" name="prev">
      <prev/>
    </do>
  </template>
  <card id="card1">
    <do type="accept">
      <go
        href="http://localhost:8080/servlet/book.servlet.Dump.Dump"
        method="post">
        <postfield name="Method" value="Yes"/>
      </go>
    </do>
    <do type="options">
      <go
        href="http://localhost:8080/servlet/book.servlet.Dump.Dump"
        method="get">
        <postfield name="Method" value="Yes"/>
      </go>
    </do>
    <p>Click ok to dump HTTP header.</p>
  </card>
</wml>
```

Figure 3.33a Example of the `method` attribute of the `<go>` element.

```
HTTP Request Method: POST
HTTP Request Method: GET
```

Figure 3.33b Screen shot progression for the `method` example.

cally, two <do> elements associate the DumpServlet with the *accept* and *options* keys, respectively. The <do> element associated with the *accept* key uses the HTTP post method, and the <do> element associated with the *options* key uses the HTTP get method. Figure 3.33b shows the method received by the DumpServlet from pressing the *accept* and *options* keys.

The <postfield> Element

The <postfield> element can appear in the content of the <go> method and is used to add option/value pairs to the content of the resulting HTTP post request. In general, the WML document in Figure 3.34a demonstrates the use of the <postfield> element. Specifically, three <postfield> elements are sent to the DumpServlet, which is

```
<wml>
  <card id="card1">
    <do type="accept">
      <go
        href="http://localhost:8080/servlet/book.servlet.Dump.Dump">
        <postfield name="Param" value="Yes"/>
        <postfield name="n1" value="v1"/>
        <postfield name="n2" value="v2"/>
      </go>
    </do>
    <p>Click ok to dump HTTP header.</p>
  </card>
</wml>
```

Figure 3.34a Example of the `<postfield>` element.

```
URL and/or Content Parameters:
  n2: v2
  n1: v1
  Param: Yes
```

Figure 3.34b Screen shot progression for the `<postfield>` example.

associated with the *accept* key. Figure 3.34b shows the option/value pairs dump associated with the example.

The <refresh> Element

The `<refresh>` element is used to refresh or establish values for WML variables. If any of the variables are displayed in the current card, the display is also refreshed. In general, Figure 3.35a demonstrates the use of the `<refresh>` element. Specifically, the example shows a deck containing two cards: card1 and card2. Upon entry into card1, a `<refresh>` element is used to establish the value of the variable `$(var): 'First time.'` When card2 is entered via the accept button on card1, a `<refresh>` element is used to modify the value of the variable. Figure 3.35b shows a screen shot progression associated with the example. When the deck is loaded, card1 is displayed as shown in panel one. Upon pressing the *accept* key, card2 appears as shown in panel two. Finally, when the *accept* key is pressed again, the default action is to pop the history stack and return to card1. Panel three shows card1 at this point.

NOTE

While the WML DTD allows empty `<refresh>` elements, it is pointless to have a `<refresh>` element without one or more `<setvar>` elements in its content.

```
<wml>
  <card id="card1">
    <onevent type="onenterforward">
      <refresh>
        <setvar name="var" value="First time"/>
      </refresh>
    </onevent>
    <do type="accept">
      <go href="#card2"/>
    </do>
    <p>
      $(var)
    </p>
  </card>
  <card id="card2">
    <onevent type="onenterforward">
      <refresh>
        <setvar name="var" value="Second time"/>
      </refresh>
    </onevent>
    <do type="accept" label="Back">
     <prev/>
    </do>
    <p>
      Press back
    </p>
   </card>
</wml>
```

Figure 3.35a Example of the `<refresh>` element.

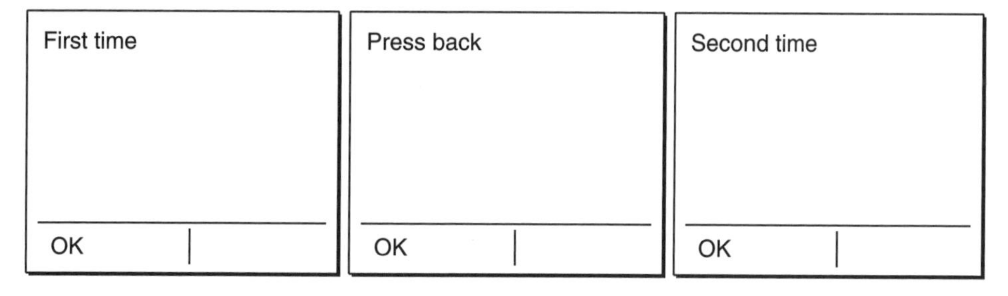

Figure 3.35b A screen shot progression for the `<refresh>` example.

The <prev> Element

The `<prev>` element is used to move the browser to the previous card on a browser's stack. Although the `<prev>` element does not take any arguments, it can have one or

```
<wml>
  <card id="card1">
    <onevent type="onenterforward">
      <refresh>
        <setvar name="var1" value="Original Value"/>
      </refresh>
    </onevent>
    <do type="accept">
      <go href="#card2"/>
    </do>
    <do type="option">
      <go href="#card3"/>
    </do>
    <p>
      $(var1)<br/>Press OK
    </p>
  </card>
  <card id="card2">
    <do type="prev" label="BACK">
      <prev/>
    </do>
    <p>
      Press BACK
    </p>
  </card>
  <card id="card3">
    <do type="prev" label="BACK">
      <prev>
        <setvar name="var1" value="New Value"/>
      </prev>
    </do>
    <p>
      Press BACK
    </p>
  </card>
</wml>
```

Figure 3.36a Example of the `<prev>` element.

more optional `<setvar>` elements as content. In general, the example shown in Figure 3.36a shows both forms of the `<prev>` element (with and without a `<setvar>` element as content). Specifically, the example shows a deck consisting of three cards: card1, card2, and card3. Upon entry into card1, a value is established for the variable `$(var1)` via the `<refresh>` element of the *onenterforward* type `<onevent>`. A `<do>` element on card1 associates card2 with the *accept* key. Furthermore, a `<do>` element associates card3 with the *option* key. Upon entry into card2, a `<do>` element associates the *back* key with the previous card via a `<prev>` element. Upon entry into card3, a `<do>` element associates the *back* key with the previous card via a `<prev>`

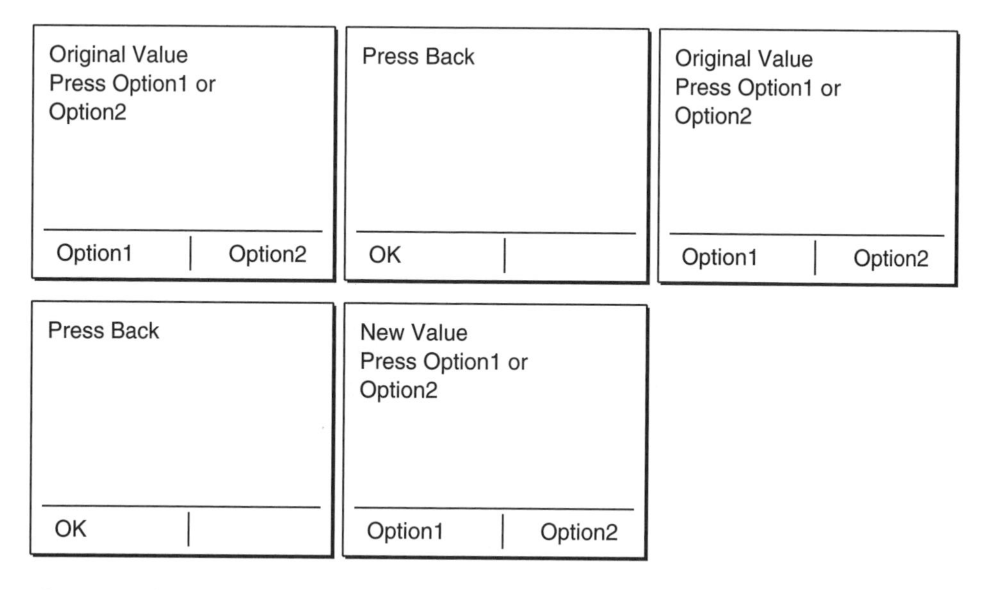

Figure 3.36b A screen shot progression for the `<prev>` example.

with a `<setvar>` element. The `<setvar>` element changes the value of `$(var1)`. Figure 3.36b shows a progression of screen shots associated with the example.

The <noop> Element

The `<noop>` element is used to prevent the normal or currently established action of a `<do>` or `<onevent>` element type attribute. As we have seen, the default behavior of the *action* key is to navigate back to the previous card on the history stack. If a developer wanted to prevent this behavior, he or she could provide a `<noop>` for the content of a `<do>` element with a type of accept.

In general, Figure 3.37a shows an example of this situation. Specifically, the example shows a deck of two cards: card1 and card2. On card2, a `<noop>` element within the content of the `<do>` element with a type of *accept* will override the default behavior, causing the *accept* key to do nothing. Figure 3.37b shows a progression of screen shots associated with the example. In panel one, card1 is displayed. Pressing the *accept* key causes card2 to be displayed due to the `<do>` element on card1. Card2 is shown in panel two. Pressing the *back* key on card2 does nothing due to the `<noop>` element.

Card Fields

The card fields category includes five elements that you can use to collect input from a user: `<select>`, `<optgroup>`, `<option>`, `<input>`, and `<fieldset>`. The following sections describe how these elements are used.

```
<wml>
  <template>
    <do type="prev">
      <prev/>
    </do>
  </template>
  <card id="card1">
    <do type="accept">
      <go href="#card2"/>
    </do>
    <p>This is card 1</p>
  </card>
  <card id="card2">
    <do type="prev">
      <noop/>
    </do>
    <p>This is card 2</p>
    <p>Press Back to go nowhere.</p>
  </card>
</wml>
```

Figure 3.37a Example of the `<prev>` element.

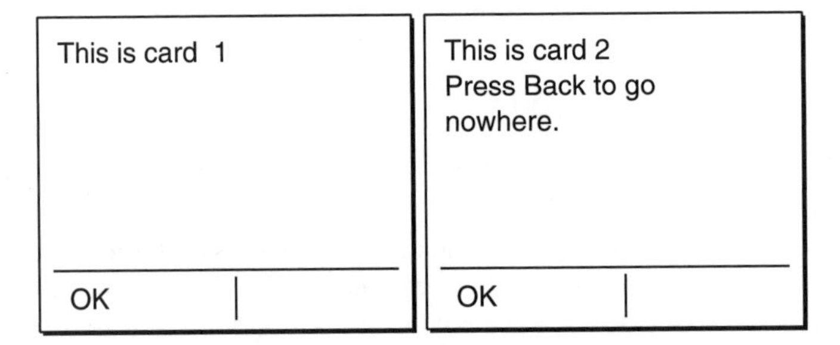

Figure 3.37b A screen shot progression for the `<prev>` example.

The <input> Element

The `<input>` element is used to prompt the user for simple text input. The element accepts the following attributes: `name`, `title`, `type`, `value`, `format`, `emptyok`, `size`, `maxlength`, and `tabindex`. The following sections describe each of these attributes.

The name Attribute

The value of the `name` attribute is the name of the WML variable designated to receive the value entered. In general, the WML document shown in Figure 3.38a demonstrates the use of the `name` attribute. Specifically, the example shows a deck consisting of one card. An `<input>` element is used to accept simple input into the variable `$(var)`. A

```
<wml>
  <card>
    <do type="accept">
      <go
        href="http://localhost:8080/servlet/book.servlet.Dump.Dump">
        <postfield name="Param" value="Yes"/>
        <postfield name="Value" value="$(var)"/>
      </go>
    </do>
    <p>Enter a value:
      <input name="var"/>
    </p>
  </card>
</wml>
```

Figure 3.38a Example of the `name` attribute for the `<input>` element.

```
URL and/or Content Parameters:
  Value: testing
  Param: Yes
```

Figure 3.38b A screen shot progression for the name attribute example.

`<do>` element associates the *accept* key with the DumpServlet. A `<postfield>` element is used to post the value of the variable `$(var)` to the Servlet. Figure 3.38b shows the parameters submitted to the DumpServlet in response to pressing the *accept* key.

The type Attribute

The `type` attribute takes one of the following two values: *text* or *password*. The implied value of the `type` attribute is *text*. Using the *text* value causes the browser to prompt for simple text input, as shown in the previous example. Using the *password* value causes the browser to mask the value as entered. In general, the WML document shown in Figure 3.39a demonstrates the use of the *password* type. Specifically, the example shows a deck consisting of one card. An `<input>` element is used to accept a password into the variable `$(var)`. A `<do>` element associates the *accept* key with the DumpServlet. A `<postfield>` element is used to post the value of the variable `$(var)` to the Servlet. Figure 3.39b shows the parameters submitted to the DumpServlet in response to pressing the *accept* key.

The title Attribute

The `title` attribute is used to establish a title for an `<input>` element. Not all browsers display the title for an `<input>` element. In general, the WML document shown in Figure 3.40a demonstrates the use of the `title` attribute. Specifically, the example shows a deck consisting of one card. On this card, the `title` attribute is used

```
<wml>
  <card>
    <do type="accept">
      <go
        href="http://localhost:8080/servlet/book.servlet.Dump.Dump">
        <postfield name="Param" value="Yes"/>
        <postfield name="value" value="$(var)"/>
      </go>
    </do>
    <p>Enter a value:
      <input name="var" type="password"/>
    </p>
  </card>
</wml>
```

Figure 3.39a Example of the `type` attribute for the `<input>` element.

```
URL and/or Content Parameters:
  Param: Yes
  value: secret
```

Figure 3.39b A screen shot progression for the `type` attribute example.

on the <input> element in order to establish the title "Name." Figure 3.40b shows the browser screen shot associated with the example.

The value Attribute

The value attribute is used to establish a default value for an <input> element. In general, the WML document shown in Figure 3.41a demonstrates the use of the value attribute. Specifically, the example shows a deck consisting of one card. On this card, the value attribute is used on the <input> element in order to establish the default value "testing." Figure 3.41b shows the browser screen shot associated with the example.

The format Attribute

The format attribute is used to establish a default value for an <input> element. The value of the format attribute is a mask through which the entered value is filtered. The mask is useful to the user in that it adjusts the keyboard of the browser. For example, if the mask indicates that a number belongs in a particular character position, not only will the input choice be limited to a numeral, but the keyboard will shift from alphanumeric to numeric mode. Table 3.2 shows what strings might appear in the value of the format attribute and how they are interpreted.

For example, if you want to limit input to three upper-case letters followed by three numbers, you would use the format attribute value AAANNN. If a single digit precedes

```
<wml>
  <card>
    <do type="accept">
      <go
        href="http://localhost:8080/servlet/book.servlet.Dump.Dump">
        <postfield name="Param" value="Yes"/>
        <postfield name="value" value="$(var)"/>
      </go>
    </do>
    <p>Enter a value:
      <input name="var" title="Name"/>
    </p>
  </card>
</wml>
```

Figure 3.40a Example of the `title` attribute for the `<input>` element.

```
URL and/or Content Parameters:
  Param: Yes
  value: secret
```

Figure 3.40b A screen shot progression for the `title` attribute example.

a format symbol, that format symbol is repeated that number of times; therefore, the previous example could be restated by using the `format` attribute value 3A3N. The * (asterisk) symbol can precede a format symbol to indicate one or more of that symbol. For example, if you wanted to specify a numeric field of one or more characters in length, you could use the `format` attribute value *N. You can use other symbol characters as formatting aids. For example, a phone number could be represented with the `format` attribute (3N)3N-4N.

In general, the WML document shown in Figure 3.42a demonstrates the use of the `format` attribute. Specifically, the example shows a deck consisting of one card. On this card, the `format` attribute is used on the `<input>` element to set up an input mask for a 5-digit zipcode. Figure 3.42b shows a screen shot associated with this example. Figure 3.42c shows the parameters submitted to the DumpServlet upon pressing the *accept* key.

The emptyok Attribute

The `emptyok` attribute is used to permit or prevent empty `<input>` element values. The `emptyok` attribute can take one of two values: *true* or *false*. The implied value is *false*. If the value is *true*, the associated `<input>` element might be empty. If the value is *false*, the associated `<input>` element must have a value before it can be passed. In general, the WML document in Figure 3.43a shows a deck containing a single card. That card contains an `<input>` element. The `<input>` element has an `emptyok` attribute

```
<wml>
  <card>
   <do type="accept">
      <go
       href="http://localhost:8080/servlet/book.servlet.Dump.Dump">
         <postfield name="Param" value="Yes"/>
         <postfield name="value" value="$(var)"/>
      </go>
    </do>
    <p>Enter a value:
      <input name="var" value="testing"/>
    </p>
  </card>
</wml>
```

Figure 3.41a Example of the `value` attribute for the `<input>` element.

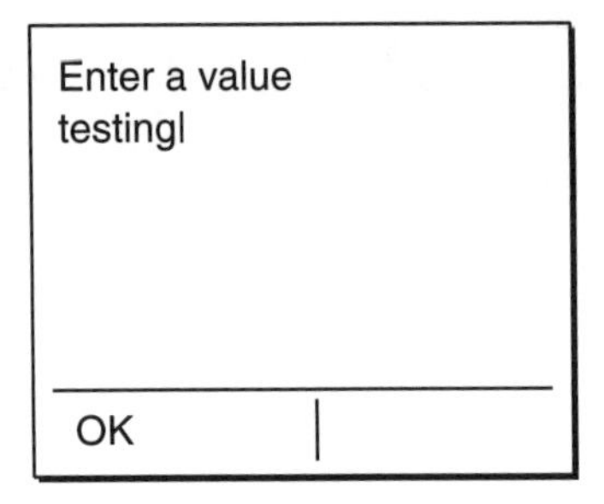

Figure 3.41b A screen shot progression for the `value` attribute example.

Table 3.2 Special Characters for the `format` Attribute

SYMBOL	INTERPRETATION
A	Upper-case letter or symbol (no numbers)
a	Lower-case letter or symbol (no numbers)
N	Number (no letters or symbols)
X	Upper-case letter, symbol, or number
x	Lower-case letter, symbol, or number
M	Upper-case letter (changeable to lower-case), symbol, or number
m	Lower-case letter (changeable to upper-case), symbol, or number

with a value of *true*. Figure 3.43b shows the screen shot associated with the example. Because input is not required, you can enter a blank value. Figure 3.43c shows the parameters submitted to the DumpServlet upon pressing the *accept* key.

```
<wml>
  <card>
    <do type="accept">
      <go
        href="http://localhost:8080/servlet/book.servlet.Dump.Dump">
        <postfield name="Param" value="Yes"/>
        <postfield name="value" value="$(var)"/>
      </go>
    </do>
    <p>Enter your zipcode
      <input name="var" format="5N"/>
    </p>
  </card>
</wml>
```

Figure 3.42a Example of the `format` attribute for the `<input>` element.

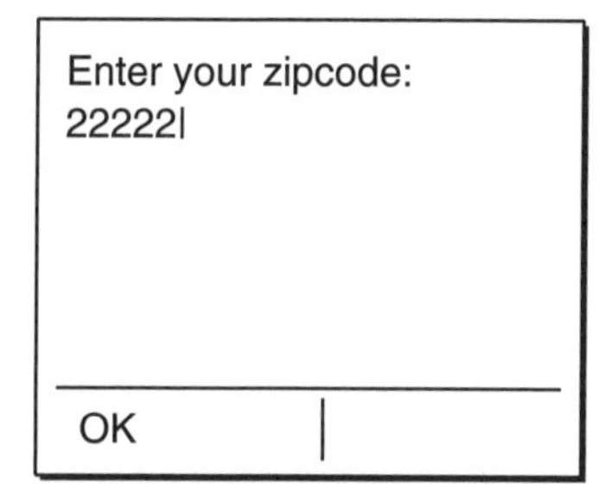

Figure 3.42b Screen shot of the `format` example.

```
URL and/or Content Parameters:
  Param: Yes
  value: 22222
```

Figure 3.42c Parameters from the `format` example.

The size Attribute

The `size` attribute is used to establish the `size` of an input field. The value of the `size` attribute is an integer. While a field might *accept* more characters than its width (refer to the `maxlength` attribute as follows), this attribute provides a means of indicating the width of the control itself. In general, Figure 3.44 shows how the `size` attribute is used.

NOTE

Implementation of the size attribute is optional for browsers. Personally, I have not run across a browser that implements this attribute.

```
<wml>
  <card>
    <do type="accept">
      <go
        href="http://localhost:8080/servlet/book.servlet.Dump.Dump">
        <postfield name="Param" value="Yes"/>
        <postfield name="value" value="$(var)"/>
      </go>
    </do>
    <p>Enter a value
      <input name="var" emptyok="true"/>
    </p>
  </card>
</wml>
```

Figure 3.43a Example of the `emptyok` attribute for the `<input>` element.

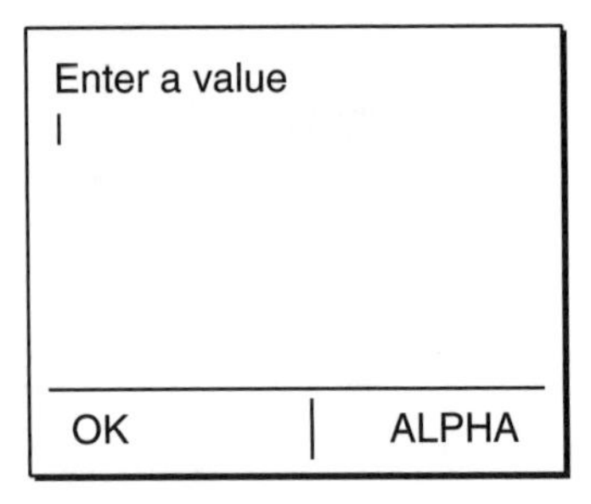

Figure 3.43b A screen shot progression for the `emptyok` example.

```
URL and/or Content Parameters:
  Param: Yes
  value:
```

Figure 3.43c Parameters from the `emptyok` example.

```
<wml>
  <card id="card1">
    <p>Date: (YYMMDD)
      <input name="var1" size="6"/>
    </p>
  </card>
</wml>
```

Figure 3.44 Example of the `size` attribute for the `<input>` element.

The maxlength Attribute

The maxlength attribute is used to establish the maximum length of an <input> element. The value of the maxlength attribute is an integer. The implied value of the attribute is an unlimited length. In general, the WML document in Figure 3.45a shows a deck containing a single card. That card contains an <input> element. The <input> element has a maxlength attribute with a value of five. Figure 3.45b shows the screen shot associated with the example. Because the length of the input field is limited to five characters, input will end once five characters have been entered. Figure 3.45c shows the parameters submitted to the DumpServlet upon pressing the *accept* key.

```
<wml>
  <card>
    <do type="accept">
      <go
        href="http://localhost:8080/servlet/book.servlet.Dump.Dump">
        <postfield name="Param" value="Yes"/>
        <postfield name="value" value="$(zipcode)"/>
      </go>
    </do>
    <p>Enter a zipcode
      <input name="zipcode" format="*N" maxlength="5"/>
    </p>
  </card>
</wml>
```

Figure 3.45a Example of the `maxlength` attribute for the `<input>` element.

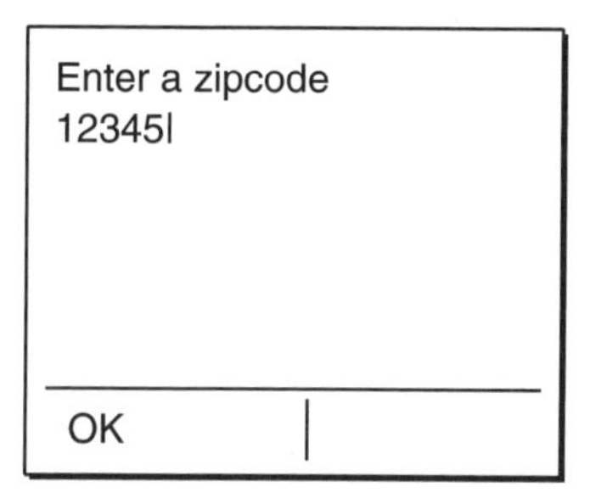

Figure 3.45b A screen shot progression for the `maxlength` attribute.

```
URL and/or Content Parameters:
  Param: Yes
  value: 12345
```

Figure 3.45c Parameters from the `maxlength` example.

The tabindex Attribute

The `tabindex` attribute is used to imply a preferred tab order associated with `<input>` and `<select>` elements on a card. The value of the `tabindex` attribute is an integer. The tab order of the fields is determined by their `tabindex`, from low to high.

TIP

The `tabindex` attribute is not an option and tends to not be implemented. I would recommend that you avoid using this attribute. If the nature of your data requires a particular order of entry, you should order your `<input>` and `<select>` elements in that order within the card.

The <select> and <option> Elements

Due to the symbiotic relationship between the `<select>` and `<option>` elements, they are best described together. The `<select>` element is used to represent a list of items from which you can select one or more items. The `<option>` elements that appear within the `<select>` element's content represent the items that are available for selection. The `<select>` and contained `<option>` elements are used primarily for one of three purposes. First, they can be used to form a menu of destinations. Second, they can be used to represent the values of a form. Third, they can be used to choose from a list of values (such as an entry in a form). We will explore examples of each in the following text.

Due to the complexity of the `<select>`/`<option>` coupling, a simple example is in order. In general, the example in Figure 3.46a shows a simple `<select>`/`<option>` example. Specifically, the example shows a deck containing one card: card1. Card1 contains a `<select>` element containing four `<option>` elements. The text appearing within the content of the option is the text that represents the item in the list. Figure 3.46b shows a screen shot of card1. Each item in the list is represented by enumerated text on the display.

While the previous example conveys the basic principles of the `<select>`/`<option>` elements, this example is of little practical use. The lack of usability stems from the fact that there is no way to determine which option the user selected. In order to take some meaningful action, the developer must choose from one of three available methods: indexing, name-value pair, or action. Using the indexing method, the developer provides a variable to receive the index of the `<option>` element selected. Using the name-value pair method, the developer provides a variable to receive the value associated with the `<option>` element selected. Using the action method, the developer provides an action for each option. We will explore each of these methods in the following sections.

Using the index Method

As stated earlier, when using the `index` method, the developer must specify a variable to receive the index of the `<option>` element selected. To provide the name of this variable, the `iname` attribute of the `<select>` element is used. The value of the `iname` attribute is the unadorned name of a WML variable.

```
<wml>
  <card id="card1">
    <p>Pick a number
      <select>
        <option>one</option>
        <option>two</option>
        <option>three</option>
        <option>four</option>
      </select>
    </p>
  </card>
</wml>
```

Figure 3.46a Example of the basic `<select>` element.

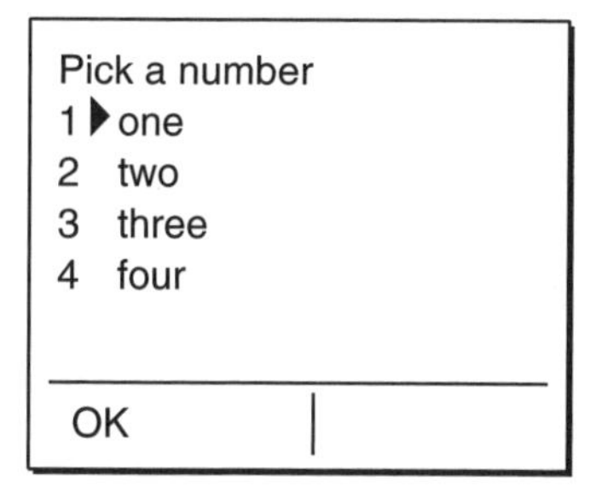

Figure 3.46b A screen shot for the basic `<select>` example.

In general, the example in Figure 3.47a shows how the `iname` attribute is used to name a select variable. Specifically, the example shows a deck that contains two cards: card1 and card2. Building on the previous example, card1 uses a `<do>` element to associate card2 with the *accept* key. Additionally, the `iname` attribute has been specified on the `<select>` element to target the variable `$(var1)` as the index variable. Once card2 is invoked (by selecting an option and pressing the *accept* key), the value of the index variable is displayed. Figure 3.47b shows the progression of screen shots that would result from selecting the second option.

NOTE

The value placed in an index variable is one based.

Using the name-value Method

As stated previously, when using the `name-value` method, the developer must specify a variable to receive the value associated with the `<option>` element selected. To provide the name of this variable, the `name` attribute of the `<select>` element is used. The value of the `name` attribute is the unadorned name of a WML variable. Additionally, when using this method, each `<option>` element should provide a value. To provide the value, the `value` attribute of the `<option>` element is used.

```
<wml>
  <card id="card1">
    <do type="accept">
      <go href="#card2"/>
    </do>
    <p>Pick a number
      <select iname="var1">
        <option>one</option>
        <option>two</option>
        <option>three</option>
        <option>four</option>
      </select>
    </p>
  </card>
  <card id="card2">
    <p>You selected option $(var1)</p>
  </card>
</wml>
```

Figure 3.47a Example of the `index` method.

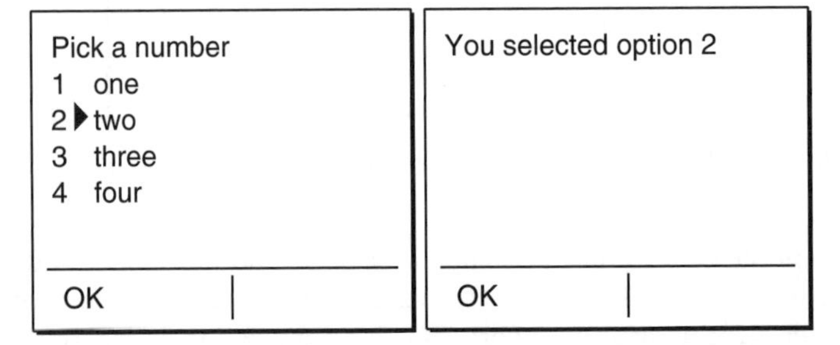

Figure 3.47b A screen shot progression for the `index` method example.

In general, the example in Figure 3.48a shows how the `name` and `value` attributes are used for the `<select>` and `<option>` elements, respectively. Specifically, the example shows a deck that contains two cards: card1 and card2. Once again, building on the original example in Figure 3.48a, card1 uses a `<do>` element to associate card2 with the *accept* key. Additionally, the `name` attribute is specified on the `<select>` element in order to target the variable `$(var1)` as the variable. Furthermore, the `value` attribute is specified on each `<option>` element in order to provide a value. Once card2 is invoked, the value of the variable appears. Figure 3.48b shows the progression of screen shots that would result from selecting the third option.

Using the action Method

When using the `action` method, you will find that each option has its own action to be performed. If the action is to invoke a simple URL based on pressing the *accept* key, the

```
<wml>
  <card id="card1">
    <do type="accept">
      <go href="#card2"/>
    </do>
    <p>Pick a number
      <select name="var1">
        <option value="one">one</option>
        <option value="two">two</option>
        <option value="three">three</option>
        <option value="four">four</option>
      </select>
    </p>
  </card>
  <card id="card2">
    <p>You selected option $(var1)</p>
  </card>
</wml>
```

Figure 3.48a Example of the `name-value` method.

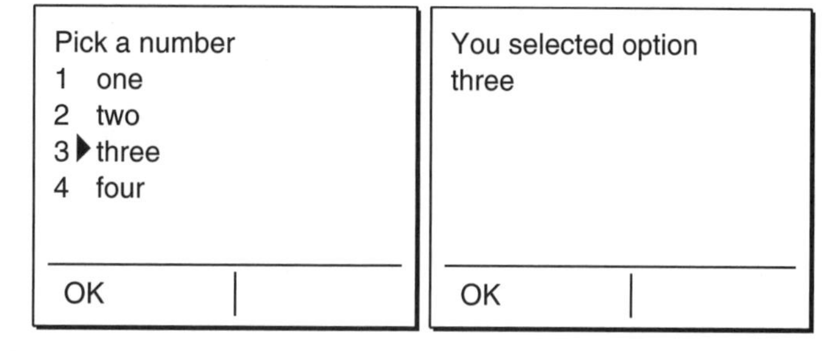

Figure 3.48b Screen shot progression for the `name-value` method example.

`onpick` attribute of the `<option>` element can be used. The value of the `onpick` attribute is a hypertext reference. If additional flexibility is required with respect to the action (and/or the action is to be associated with a key other than the *accept* key), an `<onevent>` element should be specified within the content of the `<option>`. When the `<onevent>` element is used to specify the action, the only value that can be specified for its `type` attribute is `onpick`. Otherwise, the features of the `<onevent>` element are consistent with its usage with the `<card>` and `<template>` elements (refer to the `<onevent>` element section).

In general, Figure 3.49a demonstrates both uses of the `action` method. Specifically, the example shows a deck that consists of three cards: card1, card2, and card3. On card1, a `<do>` element is used to associate card2 with the action key. Upon entering card2, the variable `$(var1)` is initialized to the default value of the example (Yes). In card two, a `<select>` element contains `<option>` elements for the possible

```
<wml>
  <card id="card1">
    <do type="accept">
      <go href="#card2"/>
    </do>
    <p>Press OK to get a question</p>
  </card>
  <card id="card2">
    <onevent type="onenterforward">
      <refresh>
        <setvar name="var1" value="Yes"/>
      </refresh>
    </onevent>
    <p>Are you sure?
      <select name="var1">
        <option onpick="#card3">Yes</option>
        <option>No
          <onevent type="onpick">
            <go href="#card3">
              <setvar name="var1" value="No"/>
            </go>
          </onevent>
        </option>
        <option>cancel
          <onevent type="onpick">
            <prev/>
          </onevent>
        </option>
      </select>
    </p>
  </card>
  <card id="card3">
    <p>You selected $(var1)</p>
  </card>
</wml>
```

Figure 3.49a Example of the `action` method.

choices (yes, no, and cancel). If the yes option is selected, control is passed to card3 via the `onpick` attribute of the `<option>` element. If the no option is selected, control is passed to card3 via the `<onevent>` element of that option. If the cancel option is selected, the previous card (card1) is backed into via the history stack. On card3, the value of the variable appears.

Figure 3.49b shows two screen shot progressions. The first progression shows the path from card1 to card2 and back to card1, which is achieved by selecting the cancel option. The second progression shows the path from card1 to card3 by selecting the yes option.

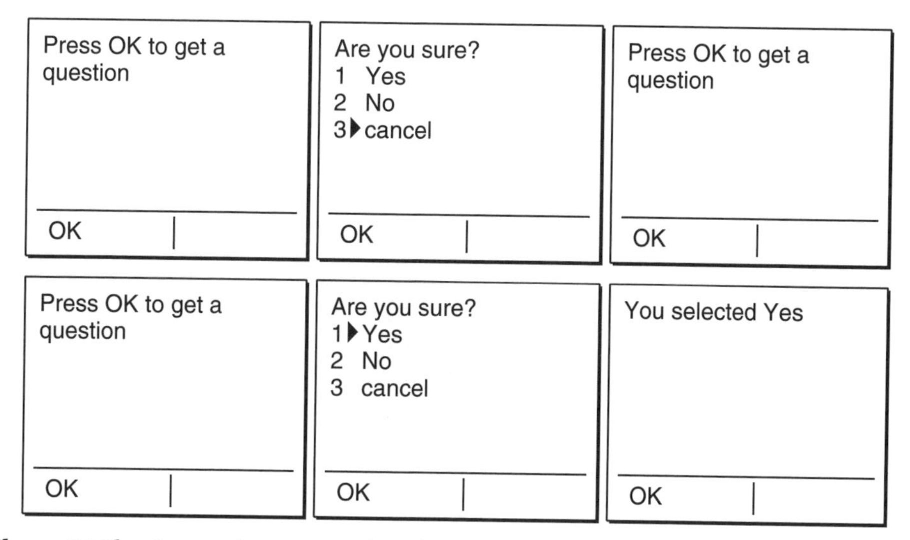

Figure 3.49b Screen shot progressions for the `action` method example.

Labeling the *Accept* Key

Sometimes, you will find it useful for one or more particular options in a select list to have an alternate label for the *accept* key. You can perform this task by using the `title` attribute. The value of the `title` attribute becomes the name of the accept button. In general, Figure 3.50a shows several examples of labeling. Specifically, the examples show a deck with one card: card1. In the example, each option in the select list is assigned a label via the `title` attribute. Figure 3.50b shows a screen shot progression where the selection cursor is positioned on each of the available choices.

Selecting a Default Option

Through your interactions with the examples presented so far, you have no doubt noted that when backing into a card with a `<select>` element, your previous selection is shown as the default. In many situations, it would be nice to programmatically set the default value. For example, if a user were modifying a value via a `<select>` element, it would be good to show the current value as the default. WML provides this capability with the `value` and `ivalue` attributes of the `<select>` element. The value of the `ivalue` attribute is the unadorned name of the WML variable that contains the index of the default item. Similarly, the value of the `value` attribute is the unadorned name of the variable that contains a value to be matched against the `value` attributes of the `<option>` elements.

Figure 3.51a shows an example that uses both methods. Specifically, the example shows a deck that contains two cards. Each card has a `<select>` element in which the default item has been selected by using either the `ivalue` or `value` attribute. Figure 3.51b shows a screen shot progression from card1 to card2 via the *accept* key.

```
<wml>
  <card id="card1">
    <p>Are you sure?
      <select>
        <option title="Go">Yes</option>
        <option title="Back">No</option>
        <option title="Back">Cancel
          <onevent type="onpick">
            <prev/>
          </onevent>
        </option>
      </select>
    </p>
  </card>
</wml>
```

Figure 3.50a Example of *accept* key labeling.

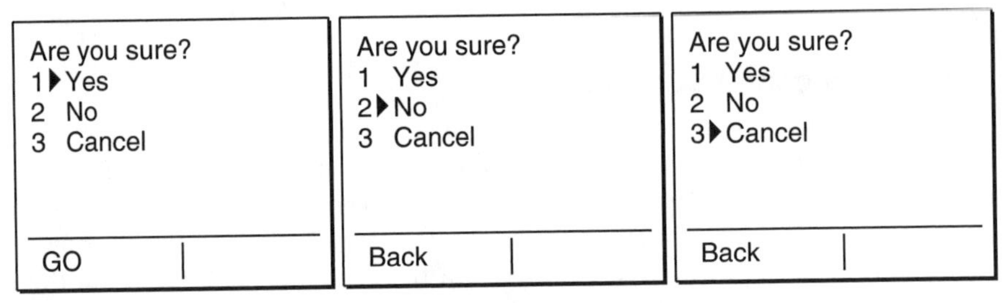

Figure 3.50b Screen shot progressions for the labeling example.

NOTE

If the variable containing the index or value for the default item is undefined or does not relate to a valid choice, the first item in the `<select>` element is used as the default. The comparison process is case sensitive.

The <select> Element

Besides the features of the `<select>` element already explored, there are three additional attributes that we must discuss: `title`, `multiple`, and `tabindex`. The following sections describe these attributes.

The title Attribute

The `title` attribute provides a means of naming a `<select>` element. The value of the attribute is the desired name. The exact meaning of the `title` attribute is browser specific; therefore, developers are encouraged to explore its behavior on the browser platforms they are targeting.

```
<wml>
  <card id="card1">
    <onevent type="onenterforward">
      <refresh>
        <setvar name="var1" value="two"/>
        <setvar name="var2" value="2"/>
      </refresh>
    </onevent>
    <do type="accept">
      <go href="#card2"/>
    </do>
    <p>Pick a number
      <select name="var1" value="var1">
        <option value="one">1</option>
        <option value="two">2</option>
        <option value="three">3</option>
      </select>
    </p>
  </card>
  <card id="card2">
    <p>Pick a color
      <select iname="var2" ivalue="var2">
        <option>Red</option>
        <option>Green</option>
        <option>Blue</option>
      </select>
    </p>
  </card>
</wml>
```

Figure 3.51a Example of selecting a default option.

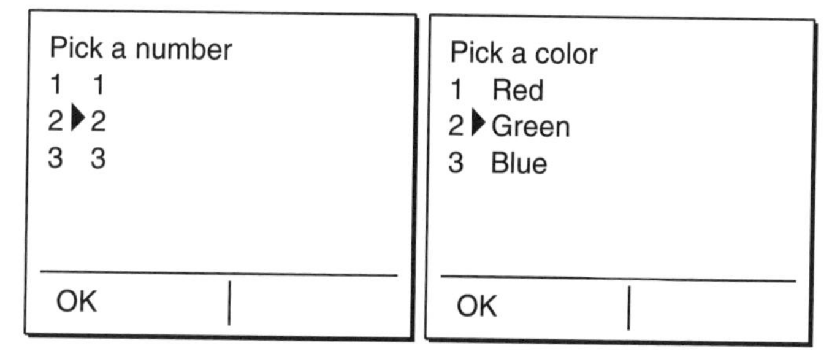

Figure 3.51b Screen shot progressions for the default option example.

The multiple Attribute

Sometimes, you will find it useful to select multiple options from a `<select>` element. WML provides a means of selecting multiple items via the `multiple` attribute.

The multiple attribute takes one of two values: *true* or *false*. The implied value is *false*. When the multiple attribute has a value of true, you can select zero, one, or many items.

In general, Figure 3.52a shows an example of multiple selections. Specifically, the example shows a deck containing two cards: card1 and card2. On card1, a <select> element employs the multiple attribute to enable many items to be selected. When control passes to card2, the results appear. Figure 3.52b shows the progression from card1 to card2 after selecting two items.

The tabindex Attribute

The tabindex attribute is used to imply a preferred tab order associated with <input> and <select> elements on a card. The value of the tabindex attribute is an integer. The tab order of the fields is determined by their tabindex (from low to high).

```
<wml>
  <card id="card1">
    <do type="accept">
      <go href="#card2"/>
    </do>
    <p>What animals do you own?
      <select multiple="true" name="animals">
        <option value="cats">cats</option>
        <option value="dogs">dogs</option>
        <option value="cows">cows</option>
      </select>
    </p>
  </card>
  <card id="card2">
    <p>You have the following animals:<br/>$(animals)</p>
  </card>
</wml>
```

Figure 3.52a Example of the multiple attribute for the <select> element.

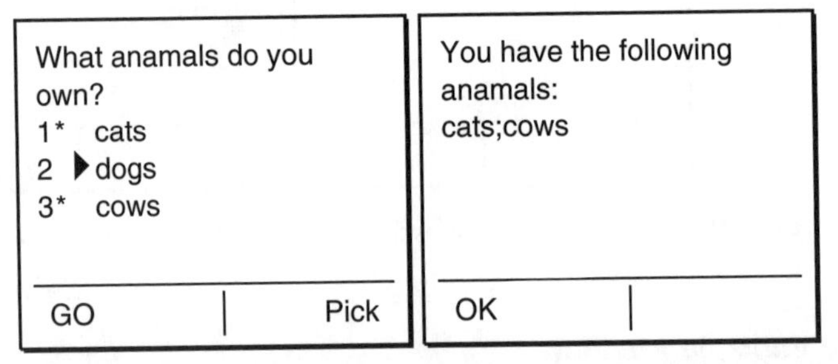

Figure 3.52b Screen shot progressions for the multiple attribute example.

TIP

The `tabindex` attribute is not an option and tends to not be implemented. I would recommend that you avoid using this attribute. If the nature of your data requires a particular order of entry, you should order your `<input>` and `<select>` elements in that order within the card.

Images

Within the images category, there is a single element: `<img>`. We discuss the `<img>` element in the following section.

The <img> Element

The `<img>` element is used to display binary images on a WAP device. While the `<img>` element takes no content, it has many attributes: `alt`, `src`, `localsrc`, `vspace`, `hspace`, `align`, `height`, and `width`. From the collection of attributes, only two are required: `alt` and `src`. The following sections describe the behavior and give example of the various attributes.

The src and alt Attributes

The `src` attribute of the `<img>` element is a text string that specifies the URL of a binary image file to be loaded. When a card containing an image is rendered on a WAP browser, the browser makes an independent HTTP get request for the binary data associated with the image (unless the image is in the browser's cache). The `src` attribute identifies the location of that resource.

Because a binary image must be loaded in a subsequent HTTP request, the WAP browser has plenty of time to render a WML card while it is waiting for the server to respond to the binary image request. Many WAP browsers will render a card placing a text string where the graphic will be placed (once the download is complete). The `alt` attribute provides you with a means of specifying the text string that will appear in place of the image during this time. Furthermore, because some WAP clients do not support graphics at all, the text string provided by the `alt` attribute serves as a replacement for the image.

TIP

Many developers like to include their company or product logo on a splash screen as part of their WAP applications. You should use the name of your product or company as the value of the `alt` attribute. By performing this action, end users will see the name of your product or application (rather than the logo image) if their client does not support images.

As stated previously, the value of the `alt` attribute is the text that replaces an image while the browser is downloading the binary data associated with the image or entirely replaces the image if the client at hand does not support images. The WML document in

```
<wml>
  <card id="card1">
    <p align="center">
      <img src=" logo.wbmp"
        alt="Logo" width="35" height="13"/>
      <br/>Product Name
      <br/>Version 1.0
    </p>
  </card>
</wml>
```

Figure 3.53a Example of the <img> element.

Figure 3.53b Screen shot progressions from the <img> example.

Figure 3.53a demonstrates the use of the `src` and `alt` attributes of the <img> element. Specifically, the example shows a WML deck consisting of a single card: card1. Figure 3.53b shows a progression of screenshots associated with the example.

The localsrc Attribute

The `localsrc` attribute is similar in function to the `src` attribute. Like the `src` attribute, the value of the `localsrc` attribute is a text string that identifies an image; however, unlike the `src` attribute, the value of the `localsrc` attribute is a simple name, rather than a URL. This simple name is used to identify a local resource stored within the browser itself. The WML specification states that the `localsrc` attribute (if present) should be considered before the `src` attribute. In other words, if the `localsrc` attribute indicates an image stored on the WML client at hand, that image will be used. If the image indicated is not supported by the client at hand, however, the `src` attribute's value will be used to download the image to be displayed.

NOTE

Local images vary between client and browser implementations. Therefore, you should back up references to local images with suitable network-based images referenced by the `src` attribute.

```
<wml>
  <card id="card1">
    <p>
      <img src="moon.wbmp"
        localsrc="moon1" alt="moon1"/>Moon
    </p>
  </card>
</wml>
```

Figure 3.54a Example of the `localsrc` attribute.

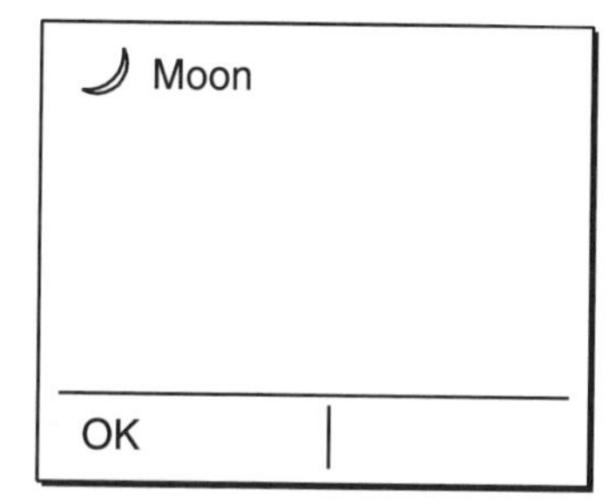

Figure 3.54b Screen shot progressions from the `localsrc` example.

In general, the WML document in Figure 3.54a demonstrates the use of the `localsrc` attribute. Specifically, the example shows a deck consisting of one card. That card contains an `<img>` element that has a `localsrc` attribute with a value of *moon1*. If *moon1* is not supported on the device, the browser will attempt to load the `src`. If the browser does not support bitmaps or both bitmaps are not supported, the text `moon1` will appear. Figure 3.54b shows screen shots associated with the example.

The vspace and hspace Attributes

The `vspace` and `hspace` attributes are used to establish padding around an image. Both attributes take an integer value representing the number of pixels to use as padding. The `vspace` attribute establishes padding on both the left and right side of the image. The `hspace` attribute establishes padding both above and below the image.

The align Attribute

The `align` attribute is used to specify vertical alignment. Horizontal alignment is determined by the paragraph within which the image appears. The `align` attribute can take one of three values: *top*, *middle*, or *bottom*. In general, Figure 3.55a shows an example of the `align` attribute. Specifically, the example contains one card that displays three lines of hyphens with images displayed using different alignments. Figure 3.55b shows a screen shot of the example.

```
<wml>
  <card id="card1">
    <p>The moon setting:<br/>
    -
    <img src="moon.wbmp" align="top" localsrc="moon1" alt=""/>
    -
    <img src="moon.wbmp" align="middle" localsrc="moon1" alt=""/>
    -
    <img src="moon.wbmp" align="bottom" localsrc="moon1" alt=""/>
    -
    </p>
  </card>
</wml>
```

Figure 3.55a Example of the `align` attribute.

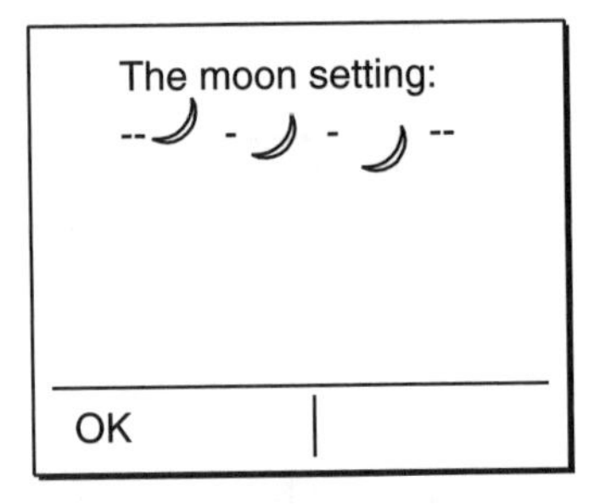

Figure 3.55b Screen shot progressions from the `align` example.

The height and width Attributes

The `height` and `width` attributes provide the client with hints about the size of the wireless bitmap. Because text is rendered on the display while images are being loaded, this information will assist the browser with laying out the bitmap. The browser is free to take whatever action is required to get the image to fit.

Anchors

Within the anchors category, there are two elements: `<a>` and `<anchor>`. The following sections describe these two elements and provide examples of their use.

The <a> Element

As we saw in some of the previous examples, the `<a>` element is used to establish a link to a WML card. The `<a>` element has one attribute: `href`. The `href` attribute is required and is used to specify a URL of the card to be loaded.

The <anchor> Element

The `<anchor>` element is similar in function to the `<a>` element; however, it provides more functionality. Where the `<a>` element only provides a means of specifying a hypertext reference in the form of a URL, the `<anchor>` element can contain the following elements: `<go>`, `<prev>`, or `<refresh>`. Besides these elements, the `<anchor>` element can also accept text and the elements that can be embedded within the `<p>` element. Figure 3.56a shows an example of the `<anchor>` element. Figure 3.56b shows a progression of screenshots associated with the example.

Tables

Within the table category, there is one base-level element: `<table>`. We discuss this `<table>` element in the following section.

The <table> Element

The `<table>`element defines a means of displaying information in a two-dimensional table. A table is built by defining `<tr>` (table rows) elements within the `<table>` element. Each `<tr>` element within a table defines a row in the table. Individual elements of table data are implemented as the content of a `<td>` (table data) element. For example, if a table contains four `<tr>` elements and each `<tr>` element contains two `<td>` elements, the resulting table will consist of four rows and two columns (for a total of eight elements).

You can modify the behavior of the `<table>` element by using three attributes: `title`, `align`, and `columns`. Of these elements, the `columns` attribute is required. The value of the `columns` element is an integer that specifies the number of `<td>` elements that will appear within each `<tr>` element.

NOTE

Some WAP browsers do not require implementation of the `columns` attribute; however, according to the specification, the attribute is required. To make sure that your table-related WML code is browser independent, make sure that you always use the attribute.

The `align` attribute is used to establish the default alignment of data within a table's column. The `align` can take one of three values: *left, center, or right.* Although not specified, the implied value of the `align` attribute is left.

The `title` attribute is used to establish a title for a table. The value of the `title` attribute is a text string to be used as the table's title.

NOTE

Some WAP browsers ignore the `title` attribute. If you find yourself in a situation where you would like to title a table, you should provide a title via a paragraph element outside the table.

```
<wml>
  <card id="card1">
    <p>Card 1<br/>Click
      <anchor>
        <go href="#card2"/>
        here
      </anchor>
      to go to card 2.
    </p>
  </card>
  <card id="card2">
    <onevent type="onenterforward">
      <refresh>
        <setvar name="var" value="1st"/>
      </refresh>
    </onevent>
    <p>Card 2 $(var) time<br/>Click
      <anchor>
        <refresh>
          <setvar name="var" value="2nd"/>
        </refresh>here</anchor>
        to reload.
    </p>
    <p>Click
      <a href="#card3">here</a>
      to go to card three.
    </p>
  </card>
  <card id="card3">
    <p>Card 3<br/>Click
      <anchor>
        <prev>
          <setvar name="var" value="3rd"/>
        </prev>
        here
    </anchor>
    to go back to card 2.</p>
  </card>
</wml>
```

Figure 3.56a Example of the `anchor` element.

Neither the <tr> nor the <td> element takes attributes. The <table> element can only contain <tr> elements as content. Likewise, the <tr> element can only contain <td> elements. The <td> element, however, can contain text, <p> elements, or any element that can appear within a <p> element (except the <table> element itself).

In general, the WML document shown in Figure 3.57a shows an example of the <table> element. Specifically, the example shows a deck consisting of one card:

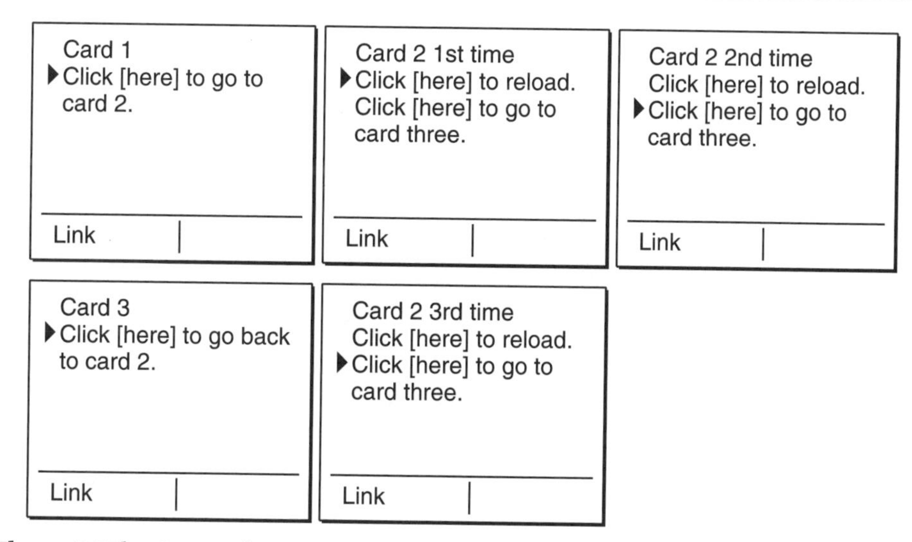

Figure 3.56b Screen shot progressions from the `anchor` example.

card1. Card1 contains a table showing the high and low temperatures for two days. Figure 3.57b shows the screen shot associated with the example.

Conclusion

In this chapter, we learned a great deal about the WML markup language. The examples shown in this chapter were to the point. In practice, your cards will contain more elements. Unless you need to break information into cards, you should put as much as you reasonably can into a single card. On a card that is being used to display information, you can use multiple `<p>` elements or one large `<p>` element with `<br/>` elements to break up the lines. On entry cards, multiple `<input>` and `<select>` elements can follow each other. The browser—according to its implementation—might either attempt to show all input-oriented elements on a single virtual screen, or it might choose to break them up over multiple screens. You will see more complex examples in the chapters to come.

TIP

I cannot stress enough the importance of getting as much experience with as many WAP browser SDKs, WAP gateways, and actual WAP devices before and during your WAP development. Implementations of browsers by different manufacturers (and, in some cases, the same manufacturer) vary widely. Implementations of WAP gateways by different manufacturers might also be different. The more experience you gain, the better your applications will be.

With regard to WML development, there are two fundamental approaches. First, write WML that is general. Do not take advantage of exotic features of WML that might

```
<wml>
  <card id="card1">
    <p>Temp:
      <table columns="3">
        <tr>
          <td>Day</td>
          <td>High</td>
          <td>Low</td>
        </tr>
        <tr>
          <td>Today</td>
          <td>70</td>
          <td>53</td>
        </tr>
        <tr>
          <td>Wed</td>
          <td>72</td>
          <td>53</td>
        </tr>
      </table>
    </p>
  </card>
</wml>
```

Figure 3.57a Example of the `<table>` element.

Temp:

Day	High	Low
Today	70	53
Wed	72	53

OK

Figure 3.57b Screen shot progressions from the `<table>` example.

behave undesirably on some browsers. This approach offers developers the best opportunity for a single code base to cover a wide range of devices. (I favor this approach in this book.) Second, write custom code for each of the popular browser and gateway combinations that you expect to encounter. While this approach requires more code and is much more complex, it produces an application that looks good on each device, rather than average on all devices.

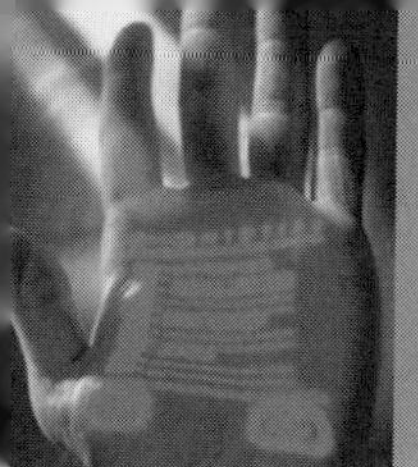

CHAPTER 4

The Java Servlet Development Kit (JSDK)

In Chapter 3, "*Wireless Markup Language (WML)*," we learned about rendering content for WAP browsers by using the WML protocol. In this chapter, we will learn about the *Java Servlet Development Kit* (JSDK). Combining these technologies will result in powerful WML-rendering servlet-engine plugins that are platform and server independent.

The JSDK is a package of Java classes and interfaces from Sun Microsystems that enables developers to add application software to a servlet engine. These applications are referred to as servlets. Because the JSDK is written in pure Java, any Java Virtual Machine should be able to execute a JSDK Servlet. Furthermore, the servlet will run on any servlet engine that supports JSDK Servlets.

The various classes and interfaces that come with the JSDK distribution reside within two packages: `java.servlet` and `java.http.servlet`. While the contents of the `java.http.servlet` package are of prime interest to the WAP Servlet developer, much of the content of this package extends functionality in the `java.servlet` package.

Because entire books have been written on JSDK (such as [1], [2], and [3]), it is clearly beyond the scope of this chapter to discuss all aspects of these packages. To bring the task to an appropriate level, we will look at the classes and methods that provide the most benefit to the WAP Servlet developer.

The Classes

The `java.servlet` package includes interfaces and classes associated with the notion of a generic servlet. The `java.http.servlet` package consists of interfaces and classes associated with servlets that plug into servlet engines. Much of the content of the java.http.servlet package extends functionality of the java.servlet package. For example, the class HttpServletRequest (from java.http.servlet) extends ServletRequest (from java.servlet). To simplify the text of this chapter, we will discuss methods in terms of the java.http.servlet package, although some are actually derived from the java.servlet package.

Five classes/interfaces are most interesting from the WAP Servlet developer's perspective: HttpServlet, HttpSession, HttpServletRequest, HttpServletResponse, and Cookie. The following sections define these classes and interfaces in greater detail.

HttpServlet

The HttpServlet class is the abstract class from which all HTTP servlets are derived. Therefore, as you develop servlets, you will be responsible for deriving your main servlet class from the HttpServlet class.

Class Definition

Figure 4.1 shows the essential methods of the HttpServlet class. These are the methods that you will need to provide in the implementation of your main servlet class.

Methods

All methods of the HttpServlet class have default implementations. Developers are therefore free to implement the methods required by their servlet, leaving all other methods undefined. The following sections describe the purpose of the method shown in Figure 4.1.

```
public abstract class HttpServlet
{
    public void init(ServletConfig config)
        throws ServletException;
    public void destroy();
    protected void doGet(HttpServletRequest req,
        HttpServletResponse res)
        throws ServletException, IOException;
    protected void doPost(HttpServletRequest req,
        HttpServletResponse res)
        throws ServletException, IOException;
    public String getServletInfo();
}
```

Figure 4.1 The HttpServlet class.

The init( ) Method

By default, a servlet engine will wait until a servlet is initially requested before loading the servlet. Many HTTP servlets provide a means for preloading servlets. In either case, after the servlet engine loads the servlet, it calls the servlet's `init()` method. The `init()` method provides a convenient place to initialize servlet data structures used during normal operation.

The servlet is invoked in response to HTTP `get` and `post` requests. While servlets are thread safe and each invocation of the servlet runs within its own thread, data members of the servlet class are available to all servlet users.

TIP

If your servlet uses resources that are shared across servlet invocations, the `init()` method is the best place to initialize them. If your servlet has no tasks to be performed at load time, you should use the default implementation of the `init()` method.

The destroy( ) Method

By default, aservlet engine will unload a servlet during its shutdown procedure. Many servlet engines provide a means of unloading a servlet. Additionally, many servlet engines provide a means of reloading a servlet. (When a servlet is reloaded, it is first unloaded.) In any case, immediately prior to unloading a servlet, the servlet engine will invoke the servlet's `destroy()` method. The `destroy()` method provides a convenient place to clean up shared resources that are used during normal operation.

TIP

If your servlet uses resources that are shared across servlet invocations, the `destroy()` method is the best place to perform any cleanup activities. If your servlet has no tasks to be performed at unload time, you should use the default implementation of the `destroy()` method.

The doGet( ) Method

The `doGet()` method of a servlet is called in response to the servlet engine receiving an HTTP get request on behalf of the servlet. This method receives two parameters: `HttpServletRequest` and `HttpServletResponse`. You can use the `HttpServletRequest` parameter to get information about the HTTP get request that caused the `doGet()` method to be invoked. The `HttpServletResponse` parameter provides the methods for replying to the browser that made the original HTTP get request.

NOTE

If you want your servlet to respond to HTTP get requests, you must provide an implementation of the `doGet()` method. If you do not provide an implementation of this method, the servlet engine will reject any HTTP get request sent to your servlet.

The doPost( ) Method

The `doPost()` method of a servlet is called in response to the servlet engine receiving an HTTP post request on behalf of the servlet. This method receives two parameters: `HttpServletRequest` and `HttpServletResponse`. You can use the `HttpServletRequest` parameter to obtain information about the HTTP get request that caused the `doPost()` method to be invoked. The `HttpServletResponse` parameter provides the methods for replying to the browser that made the original HTTP get request.

NOTE

If you want your servlet to respond to HTTP post requests, you must provide an implementation of the `doPost()` method. If you do not provide an implementation of this method, the servlet engine will reject any HTTP post request sent to your servlet.

TIP

As we will discuss shortly, name-value pairs (both on the request line and within the content of a post request) are accessible by using the `getParameter()` method of the `HttpServletRequest` interface. Therefore, unless there is a good reason to distinguish between HTTP get and post requests, you should implement your `doGet()` method so that it calls your `doPost()` method. Put all get and post-related logic within the `doPost()` method.

The getServletInfo( ) Method

Many HTTP servlet engines provide a means for accessing information about servlets that are loaded. The `getServletInfo()` method provides a means of returning a text string for this purpose. The default implementation returns no information.

TIP

While there is no standard format for the text string returned by `getServletInfo()`, it is useful to include a readable name, a version number, and either the company or author's name.

The HTTPSession Class

From the perspective of the HTTP protocol, each HTTP request (get or post, in the case of WAP) is an independent, stateless transaction. In other words, there is no mechanism built into HTTP that enables a servlet engine to remember the state or past transactions processed for a given client. Many Web applications need to provide the end user with a notion of a session. For example, a shopping-cart application must have the capability to remember from request to request what items have been placed in the shopping cart. By using a well-known URL parameter or an HTTP cookie, an application can provide itself with an ID that enables it to implement the notion of an end-user session. Prior to servlets, developers had to invent and implement their own methodology for accomplishing this task.

The `HttpSession` object provides a means of tracking session-oriented data by using a common and consistent methodology. As shown next (refer to `HttpServletRequest`), the `HttpServletRequest` interface provides a method for creating and/or obtaining an already created `HttpSession` object. The primary benefit of the `HttpSession` object is the capability to create and maintain session-oriented values associated with a client over multiple HTTP requests.

Class Definition

Figure 4.2 shows the essential methods of the `HttpSession` interface. We show the complete implementation of the interface in Appendix B.

Methods

The following sections describe the methods of the `HttpSession` class that are most useful to the WAP Servlet developer.

The invalidate( ) Method

The `invalidate()` method is used to terminate an existing session. For example, in a shopping-cart application, the notion of a session might begin when adding the first product to the shopping cart and end at checkout time. In this case, the `invalidate()` method would be invoked during the checkout process.

The putValue( ) Method

As stated earlier, the primary advantage of maintaining a client session is the storage and retrieval of session-oriented values across the many HTTP requests. The `putValue()` method enables the developer to store any object derived from `java.lang.Object` by name. Any value associated with the session that uses this method will be available for this request or for any other request associated with the session in the future.

Again, using the example of a shopping cart, one could store a vector of store items under the name *cart*. The invocation of the `putValue()` method would associate the

```
public interface HttpSession {
    public void invalidate ();
    public void putValue (String name, Object value);
    public Object getValue (String name);
    public void removeValue (String name);
    public String [] getValueNames ();
    public boolean isNew ();
}
```

Figure 4.2 The `HttpSession` interface.

vector with the name. On future requests associated with the session, the same vector could be accessed by using the assigned name.

The getValue( ) Method

The `getValue()` method enables the servlet to retrieve any previously stored session value by name. Prior to using the `getValue()` method, a prior call to `putValue()` must have been made for the given value name; otherwise, *null* is returned.

Building on the shopping cart example, as additional store items are purchased, the shopping cart must be updated. Recalling that the cart is maintained in a vector associated with the session value *cart*, a call is made to the `getValue()` method in order to retrieve the vector. Once the vector is retrieved, new shopping cart items can be added —and processing can continue. The servlet does not need to call `putValue()` a second time, because the vector will still be associated with the session value *cart*.

The removeValue( ) Method

The `removeValue()` method enables the servlet to remove an existing session value by name. Prior to removing a value, a value must have been previously established via the `putValue()` method.

Once again, building on the shopping cart example, if the end user elects to remove all items from the shopping cart prior to checking out, the servlet could simply remove the session value called *cart* in response to the request.

The getValueNames( ) Method

The `getValueNames()`method returns an enumeration of the values associated with the session at hand. This method is particularly useful for looping through the values. For example, the author of a servlet might want to remove all values associated with a session at the logical conclusion of the session. Alternatively, the author of a servlet might want to display all values associated with a servlet as a debugging aid.

The isNew( ) Method

The `isNew()` method returns a Boolean value. If the value is *true*, the session is new; otherwise, the value is *false*. A session is new if the current request (the one in progress) was the request in which the session was created. This value is useful for determining whether initialization needs to be performed on the session values.

HttpServletRequest

The `HttpServletRequest` interface represents an object generated by the servlet engine when an HTTP get or post request is received. When the object is created, it is initialized with information taken from the HTTP request. This object is passed as a parameter to the `doGet()` and `doPost()` methods of the servlet. Thus, the `HttpServletRequest` parameter serves as a means of getting information associated with the HTTP request.

```
public abstract class HttpServletRequest
{
    public ServletInputStream getInputStream() throws IOException;
    public String getParameter(String name);
    public Enumeration getParameterNames();
    public Cookie[] getCookies();
    public String getHeader(String name);
    public int getIntHeader(String name);
    public long getDateHeader(String name);
    public Enumeration getHeaderNames();
    public HttpSession getSession (boolean create);
}
```

Figure 4.3 The `HttpServletRequest` interface.

Class Definition

Figure 4.3 shows the essential methods of the `HttpServletRequest` and `Servlet-Request` interfaces from the perspective of the `HttpServletRequest` interface. The complete implementation of both interfaces appears in Appendix B.

Methods

The following sections describe the methods of the `HttpServletRequest` class that are most useful to the WAP Servlet developer.

The getParameter( ) Method

The `getParameter()` method returns the value of a parameter as a text string. The name argument of the `getParameter()` method is the name part of a name-value pair appearing in the request or the content of the request. The text string returned is the value part of that name-value pair.

The getParameterNames( ) Method

The `getParameterNames()` method returns an enumerated list of parameter names. This list consists of the name part of all name-value pairs appearing in the request or in the content of the request. This method is particularly useful when the servlet must loop on all parameters associated with a request (for example, as a debugging aid).

The getCookies( ) Method

One or more optional HTTP cookies can accompany any given HTTP request. For each cookie received in conjunction with an HTTP request, the servlet engine creates a `Cookie` object. The `getCookies()` method provides a means for accessing the cookies associated with an HTTP request by returning an array of associated `Cookie` objects. If no cookies are associated with a request, this method returns a *null* value.

The getHeader( ) Method

HTTP request headers are expressed as name-value pairs. The `getHeader()` method provides a means of returning string representations of a named HTTP header value. Most header values that are of interest to the WAP Servlet developer are accessible via convenience methods; however, the developer can use this method to access text values for headers that do not have a convenience method. Developers should not use this method when the servlet expects the value associated with an HTTP header to be an integer or a date. In the case of an integer value, the `getIntHeader()` method should be used. In the case of a date value, the `getDateHeader()` method should be used.

The getIntHeader( ) Method

The `getIntHeader()` method provides a means of returning the integer representation of a named HTTP header value. This method should not be used when the servlet expects the value associated with an HTTP header to be a text string or a date. In the case of a text value, the `getHeader()` method should be used. In the case of a date value, the `getDateHeader()` method should be used.

The getDateHeader( ) Method

The `getDateHeader()` method provides a means of returning the date representation of a named HTTP header value. This method should not be used when the servlet expects the value associated with an HTTP header to be a text string or an integer. In the case of a text value, the `getHeader()` method should be used. In the case of an integer value, the `getIntHeader()` method should be used.

Getting the value of a date header as a string by using the `getHeader()` method is completely acceptable; however, using the `getDateHeader()` method enables you to deal with a date header as a `java.util.date` object. This convenience is highly recommended.

The getHeaderNames( ) Method

The `getHeaderNames()` method returns an enumerated list of header value names. This list consists of the name part of all name-value pairs appearing within the HTTP request header. This method is particularly useful when the servlet must loop on all headers associated with a request (for example, as a debugging aid).

The getSession( ) Method

The `HttpSession` object provides a means of tracking session-oriented data across multiple HTTP requests. The `getSession()` method returns the `HttpSession` object associated with the client, provided that the object exists. If the parameter has a value of *true*, the session object will be created if it does not already exist.

HttpServletResponse

The HttpServletResponse interface represents an object generated by the servlet engine when an HTTP get or post request is received. This object is passed as a parameter to the doGet() and doPost() methods of the servlet. The HttpServletResponse parameter provides a means for the servlet to respond to the requesting client.

Class Definition

Figure 4.4 shows the essential methods of the HttpServletResponse and ServletResponse interfaces from perspective of the HttpServletResponse interface. The complete implementation of both interfaces appears in Appendix B.

Methods

The setContentType( ) Method

The setContentType() method is used to establish the *Multi-Purpose Internet Mail Extensions* (MIME) encoding type of the content portion of the HTTP response. For the purposes of a WAP Servlet, the MIME encoding type will normally be the following: text/vnd.wap.wml.

The getWriter( ) Method

The getWriter() method returns a PrintWriter object. This PrintWriter object is used for printing the content portion of the HTTP response. In the case of a WAP Servlet, the content portion will normally be a WML document consisting of printable characters. The PrintWriter object returned provides an excellent vehicle for accomplishing this task.

```
public interface HttpServletResponse {
    public void setContentType(String type);
    public PrintWriter getWriter() throws IOException;
    public void addCookie(Cookie cookie);
    public void setHeader(String name, String value);
    public void setIntHeader(String name, int value);
    public void setDateHeader(String name, long date);
    public void sendError(int sc, String msg) throws IOException;
    public void sendError(int sc) throws IOException;
    public void sendRedirect(String location) throws IOException;
    public String encodeUrl (String url);
    public String encodeRedirectUrl (String url);
}
```

Figure 4.4 The HttpServletResponse interface.

TIP

When you have finished printing the content of the HTTP response to the `PrintWriter` object obtained by using the `getWriter()` method, it is your responsibility as a servlet developer to close `PrintWriter`. Failure to do so could result in problems with server implementations that expect you to close the application. While some server implementations (such as the one provided with this book) will flush and close `PrintWriter` for you, others will not.

The addCookie( ) Method

The `addCookie()` method provides a means of placing a cookie into the HTTP response. This method takes a `Cookie` object as a parameter (refer to the section concerning the `Cookie` class).

TIP

The session feature of the JSDK is intended for tracking a session's values. This feature is not intended to track an end user or client across multiple sessions. If your servlet needs to track an end user or client data across multiple sessions, you should establish a cookie for that purpose. Due to privacy concerns, it is considered extremely poor form to store data in a cookie. To accomplish multi-session tracking, your servlet should store information on the server and create a unique ID to access that information. By storing the ID in an HTTP cookie, you can locate the data without placing it directly into the cookie.

The setHeader( ) Method

From time to time, a servlet developers might need to set an HTTP header directly. They can use the `setHeader()` method to accomplish this task. Most HTTP headers that are of interest to a WAP Servlet developer are set automatically or via another convenience method of the `HttpServletResponse` interface. For example, the following lines of code have the same effect:

```
setContentType("text/vnd.wap.wml");
setHeader("Content-Type", "text/vnd.wap.wml");
```

TIP

When possible, you should use the various convenience methods to set HTTP header values. They are easier to remember and make code less busy. If you find yourself needing to set a header directly, refer to the complete implementation of the `HttpServletResponse` interface in Appendix B to determine whether a convenience method fits the bill.

The setIntHeader( ) Method

The `setIntHeader()` method enables the servlet developer to establish an integer value for an HTTP header. All HTTP headers are encoded as text; however, this method provides a convenient means of setting a header that has a numeric value. Otherwise, this method is similar to the `setHeader()` method.

The setDateHeader( ) Method

The `setDateHeader()` method enables the servlet developer to establish a date value for an HTTP header. This method is useful in that it converts a `java.util.Date` object into the correct HTTP header format of a date. Otherwise, this method is similar to the `setHeader()` method.

The sendError( ) Method

The two `sendError()` methods are used to send an HTTP error response to the requesting client. Both forms of the method take an HTTP error code. One method enables users to specify their own error text, while the other uses default error text (from the server implementation of the JSDK) associated with the error code. When `sendError()` is called, the error response is immediately sent; therefore, no additional action should be taken by the servlet with regard to the `HttpServletResponse` object.

NOTE

The `sendError()` methods automatically place an HTML message into the HTTP content of the error response. Most WAP browser implementations appear to ignore the content portion of an HTTP error response. Furthermore, if an exception is thrown from an `HttpServlet` object's `doGet()` or `doPost()` method, an HTTP error response of this form is sent. If your project schedule permits an extra degree of care, you might consider catching these exceptions and processing errors in general by sending a WML response to the end user. This procedure will make for a more user-friendly application.

The sendRedirect( ) Method

Periodically, a servlet must redirect a client to another URL. The `sendRedirect()` method provides a means of sending an HTTP redirect. Upon calling this method, the redirect is immediately sent; thus, no additional action should be taken by the servlet with regard to the `HttpServletResponse` object. The method takes a text URL as a parameter. This URL represents the WAP deck or card to which the browser is being redirected.

The encodeUrl( ) Method

By default, the session capability of the JSDK is implemented by using cookies. This mechanism works well unless a browser does not support cookies or unless the end user has disabled cookies. In both of these cases, an alternative technique is required for getting the session ID back to the servlet on subsequent invocations. The `encodeUrl()` method provides this functionality. This method takes a URL as a parameter and returns an identical URL with the session ID information encoded as a name-value pair.

NOTE

In order for your servlet to work with a wide range of client/browser implementations, you should use the `encodeUrl()` method to encode all URLs associated with a session.

The encodeRedirectUrl( ) Method

Because the `sendRedirect()` method takes a URL as a parameter, the `encode-RedirectUrl` provides a means of encoding the URL with the session ID. This method is similar in function to the `encodeUrl()` method.

The Cookie Class

Because HTTP is a stateless protocol, it became important in the early days of HTTP to provide a means of storing server information on a client for future reference. These pieces of data are known as *cookies*. The `Cookie` class represents HTTP cookies with respect to the JSDK. As cookies are parsed from an HTTP request, they are placed in `Cookie` objects that are stored in the `HttpServletRequest` object. When a servlet seeks to establish a cookie on a client, it must create a `Cookie` object and add it to the `HttpServletResponse` object. For additional information about cookies, refer to [RFC 2109].

Class Definition

Figure 4.5 shows the essential methods of the `Cookie` class. The complete implementation of the class appears in Appendix B.

Methods

The following sections describe the methods of the Cookie class that are most useful to the WAP Servlet developer.

The Class Constructor

The `Cookie` class constructor takes two string parameters: `strName` and `strValue`. Because HTTP cookies are essentially name-value pairs, the name and the value come from these parameters. Cookies are also associated with a domain and a path. The

```
public class Cookie Implements java.lang.Clonable
{
    public Cookie (String strName, String strValue);
    public String getName();
    public String getValue();
    public int getVersion();
    public void setDomain(String strDomain);
    public void setMaxAge(int nExpiry);
    public void setPath(String strURL);
    public void setSecure(boolean bSecure);
    public void setVersion(int nVersion);
}
```

Figure 4.5 The `Cookie` class.

domain refers to the server domain to which the cookie should be sent. The path refers to the HTTP resource path on a server within that domain to which the cookie should be sent. Once a cookie is established on a browser, any HTTP request sent to a server and path that match the server and path of the cookie will result in the browser including the cookie with the request. By using cookies, the server can prearrange for a browser to send it information that will enable it to find locally stored information about the client (such as an end-user's profile).

NOTE

For privacy reasons, only IDs that are meaningful to the originating servlet should be stored in a cookie. You should never store personal information in a cookie (such as your name, address, phone number, or credit card information).

The getName( ) Method

The `getName()` method returns the name part of the cookie's name-value pair. This method is most useful when looking for a cookie that has been received in a request. Recalling that the `getCookies()` method of the `HttpServletRequest` object returns an array of `Cookie` objects, you must loop on this array and check the name of each cookie when looking for a particular entry.

The getValue( ) Method

The `getValue()` method returns the value part of the cookie's name-value pair. Once a desired cookie has been located, this method can be used to get its value.

The getVersion( ) Method

Cookies have a version number. This version number describes the format of the cookie header in the HTTP request. Version 0 is the original Netscape protocol. Version 1 is an experimental protocol.

The setDomain( ) Method

When establishing a cookie to be sent to a client, you can use the `setDomain()` method to specify a domain other than the default. The default value for the domain is the domain of the originating server.

The setMaxAge( ) Method

When establishing a cookie to be sent to a client, a self-destruct timer can be established on the cookie. Once the cookie has expired on the client, it will no longer be sent to a server in an HTTP request.

NOTE

As you might imagine, keeping data structures that describe end users who have visited your servlet indefinitely might put a strain on system resources. Rather than keeping such information around for extended periods of time, you might want to periodically clean up old entries. Let's say that you are implementing a shopping-cart application and want to provide the ability for end users to resume a shopping

session at some point in the future (beyond the current session). You might elect to keep the customers' carts in your server's database and establish cookies for retrieving those records. You might also elect to keep a pending shopping cart for only six months, cleaning up the local copy after that period of time. In this case, it would be nice to set the cookie so that it can be freed on the end user's client at that time. Remember, WAP clients have limited storage capabilities, and cookies take up space.

The setPath( ) Method

When establishing a cookie to send to a client, you can use the `setPath()` method to establish a path other than the default. The default path is the one used when establishing a cookie.

The path is part of the information that is considered when a client determines whether a cookie should be sent with a request. If the path part of the URL in a request being sent is at or below the level of the path established in the cookie, the cookie would be included in the request. If the `setPath()` method is used to establish a path of / (forward slash) for a given cookie, that cookie will match any requested path.

The setSecure( ) Method

If the value of a cookie contains sensitive information (and it really should not), the originating servlet can specify that the cookie should only be sent over a secure connection. You can use the `setSecure()` method to accomplish this task. If the value of the `bSecure` parameter is *true*, when the cookie is established on the client, it will only be sent on requests matching the domain and path (provided that the connection is secure).

The setVersion( ) Method

The `setVersion()` method is used to establish the version number of a cookie (refer to `getVersion()`, which was described earlier).

A Simple Servlet Example

You might recall from Chapter 3, "*Wireless Markup Language (WML)*," that a servlet was used to dump HTTP header information. This section examines that servlet in detail. This servlet is a good starting point for two reasons. First, it exercises most of the interfaces, classes, and methods previously documented within this chapter. Second, it does not have a great deal of WML; thus, we can concentrate on the JSDK content.

Requirements

The DumpServlet was written for this book and has two basic requirements. First, it should demonstrate the use of as many of the essential JSDK methods as reasonable in order to serve as a simple example of a JSDK Servlet. Second, it should be capable of dumping various parts of the HTTP requests in order to support the documentation provided in Chapter 3.

The Implementation

The following sections break this relatively simple servlet into functional blocks and discuss the use of the various JSDK methods.

Import Section

As with all Java programs, JSDK Servlets have a package declaration as well as import statements. Figure 4.6 shows the import section of the DumpServlet.

All examples shown in this book reside within the book package. Because this servlet is defined in this chapter and its name is `DumpServlet`, its fully qualified package name is `book.chap04.DumpServlet`. The source code for the example is located on the companion CD-ROM in the directory `/code/book/chap04/DumpServlet`.

Referring to Figure 4.6, the servlet uses each of the classes and interfaces defined previously. Besides the *java* package classes, there are two additional javax package classes that we did not discuss earlier: `ServletException` and `ServletConfig`. Many `HttpServlet` class methods throw the `ServletException`; hence, it was imported. We will discuss the `ServletConfig` class next.

The DumpServlet Class Declaration

The class declaration for the `DumpServlet` appears as follows:

```
public class DumpServlet extends HttpServlet
```

```
package book.chap04.DumpServlet;

import java.io.IOException;
import java.io.PrintWriter;
import java.util.Enumeration;

import javax.servlet.ServletException;
import javax.servlet.ServletConfig;
import javax.servlet.http.Cookie;
import javax.servlet.http.HttpServlet;
import javax.servlet.http.HttpServletRequest;
import javax.servlet.http.HttpServletResponse;
import javax.servlet.http.HttpSession;
```

Figure 4.6 Package and import statements.

As with all JSDK Servlets, the DumpServlet extends the HttpServlet class. When the server loads and interacts with a JSDK Servlet, it refers to it through the HttpServlet class. If the servlet class was not an extension of the HttpServlet class, the servlet engine would be unable to load it.

Class Member Declarations

Figure 4.7 shows the class member declarations of the DumpServlet.

Refer to Figure 4.7. In this figure, there is a single static class member: sm_nMaxSessionHits. The sm_nMaxSessionHits member defines the number of times that a client can send a request to the servlet (hit it) via a given session. Once the maximum number of hits has been reached, the servlet session is invalidated. Because this value is a class member, it is shared across all instances of the DumpServlet.

NOTE

Hit counting has nothing to do with the dumping of interesting HTTP request information. Hit counting is a totally fabricated concept that enables us to talk about some of the JSDK methods that we would otherwise be unable to talk about.

The init( ) Method

The init() method is used to initialize any data members that need to be initialized. Figure 4.8 shows the implementation of the init() method.

The init() method is defined in the HttpServlet class; therefore, the server is able to call this method. Immediately after the server loads a servlet, the init() method is invoked. Refer to Figure 4.8. When the init() method is defined in a servlet's implementation, the ServletConfig argument passed in line 001 must be passed to the HttpServlet class's init() method, as shown in line 003. The functionality of the

```
protected final static int sm_nMaxSessionHits = 3;
```

Figure 4.7 Class member declarations.

```
001 public void init(ServletConfig config) throws ServletException
002 {
003  super.init(config);
004 }
```

Figure 4.8 The init() method.

ServletConfig class is not of particular interest to the WAP Servlet developer; however, this aspect of servlet initialization is essential for proper servlet behavior.

NOTE

In the DumpServlet, there really is no need for an init() method. On the other hand, if the servlet needed to read persistent information stored in a file or local database, the init() method would be the ideal place to accomplish this task.

The destroy( ) Method

The destroy() method is used to clean up the servlet before it is unloaded. Figure 4.9 shows the implementation of the destroy() method.

The destroy() method is defined in the HttpServlet class; therefore, the server is able to call the method. Immediately before the server unloads a servlet, the server calls the destroy() method. Referring to Figure 4.9, we see that the HttpServlet class's version of the destroy() method must be called.

NOTE

In the DumpServlet, there really is no need for a destroy() method. On the other hand, if the servlet needed to gracefully close a file or a database, the destroy() method would be the ideal method to use.

The getServletInfo( ) Method

The getServletInfo() method is used to return information about the servlet to the server. Figure 4.10 shows the implementation of the getServletInfo() method.

```
001 public void destroy()
002 {
003  super.destroy();
004 }
```

Figure 4.9 The destroy() method.

```
public String getServletInfo()
{
    return "DumpServlet by John Cook Copyright (c) 2000";
}
```

Figure 4.10 The getServletInfo() method.

The `getServletInfo()` method is defined in the `HttpServlet` class; therefore, the server can call this method in order to obtain the text string provided by the servlet.

NOTE

As shown in Figure 4.10, the `getServletInfo()` method should return meaningful information (the name of the servlet, the name of the owner, the copyright, and so on).

The doGet( ) Method

When a browser sends an HTTP request to the DumpServlet, the servlet engine invokes the `doGet()` method of the DumpServlet. Figure 4.11 shows the implementation of the `doGet()` method.

NOTE

From the point of view of a servlet, there is not much difference between an HTTP get and post request. A post request is a little more flexible than a get request in that parameters can be included within the content. In my servlet development, I tend to forward all get requests to my servlet's `doPost()` method.

The doPost( ) Method

Because WAP browsers use HTTP get and post requests and all get requests in the DumpServlet are forward to `doPost()`, this method becomes the focal point for HTTP requests sent to the DumpServlet. Figure 4.12 shows the implementation of the `doPost()` method.

As discussed previously, the `doPost()` request receives an `HttpServletRequest` and `HttpServletResponse` object from the server. Referring to Figure 4.12, you will see that the first order of business is to establish a session for the client. Because the DumpServlet has a session of sorts, the `getSession()` method is called on the request object in order to get the browser's current session (line 004). Because the `bCreate` parameter has a value of *true*, a new session will be created and returned in the

```
public void doGet(HttpServletRequest req, HttpServletResponse resp)
  throws ServletException, IOException
{
    doPost(req,resp);
}
```

Figure 4.11 The `doGet()` method.

```
001  public void doPost(HttpServletRequest req, HttpServletResponse resp)
002     throws ServletException, IOException
003  {
004     HttpSession session = req.getSession(true);

005     if (isRequested("Method", req))
006     {
007       doDumpMethod(req);
008     }

009     if (isRequested("OVP", req))
010     {
011       doDumpOVP(req);
012     }

013     if (isRequested("Param", req))
014     {
015       doDumpParam(req);
016     }

018     boolean bTerminated = false;

019     if (isRequested("Stats", req))
020     {
021       doHandleStats(session, req, resp);
022     }

023     sendWMLResponse(resp);
024  }
```

Figure 4.12 The `doPost()` method.

event that a session does not already exist for the requesting browser. If, however, the browser sends a valid ID associated with an active session, the server will return the session associated with that ID.

The DumpServlet has four parameters that can be passed via the HTTP request: *Method*, *OVP*, *Param*, and *Stats*. On lines 005–022, the `isRequested()` method is called for each parameter (refer to the `isRequested` method section as follows). This method will return *true* if the named parameter was specified in the request with a value of *Yes*; otherwise, *false* is returned. If the value returned is *true*, in each case the method will call the appropriate dump method (refer to the section concerning individual dump methods as follows). The `doHandleStats()` method will return *true* if the session is terminated (line 021). Finally, the `sendWMLResponse()` method is called to send the WML response to the requesting browser. Refer to the `sendWMLReponse` method section as follows.

The isRequested( ) Method

As discussed previously, the `isRequested()` method returns one of two values: *true* or *false*. If the parameter passed to the method (via the `strParamName` variable) exists and has a value of *Yes*, the method will return *true*; otherwise, *false* is returned. Figure 4.13 shows the implementation of the `isRequested()` method.

Referring to Figure 4.13, you can see that the `getParameter()` method is called in order to obtain the value of the named parameter (line 004). If the named parameter appeared in either the request's URL or its content, the value associated with the parameter will be returned; otherwise, *null* is returned. If the parameter was specified but no value was provided, an empty string results. As shown on lines 005–008, the return value is set to *true* if the parameter was specified with a value of *Yes*.

The doDumpMethod( ) Method

One of the methods of the `HttpServletRequest` interface not previously documented is `getMethod()`. While it is useful in an HTTP DumpServlet and for debugging, this method is not particularly useful in a typical production servlet. Figure 4.14 shows the implementation of the `doDumpMethod()` method.

```
001 protected boolean isRequested(String strParamName)
002 {
003    boolean bRetval = false;

004    String m_strParamValue = m_request.getParameter(strParamName);

005    if (m_strParamValue != null)
006    {
007       bRetval = m_strParamValue.equalsIgnoreCase("Yes");
008    }

009    return bRetval;
010 }
```

Figure 4.13 The `isRequested()` method.

```
001 protected void doDumpMethod()
002 {
003    System.out.println("HTTP Request Method: " + m_request.getMethod());
004 }
```

Figure 4.14 The `doDumpMethod()` method.

Refer to Figure 4.14. The `getMethod()` method is called in order to obtain the HTTP request method (get or post), as in line 003.

The doDumpOVP( ) Method

The `doDumpOVP()` method dumps the HTTP header options. Figure 4.15 shows the implementation of the `doDumpOVP()` method.

Referring to Figure 4.15, you can see that the `getHeaderNames()` method is called on the request object in order to obtain an enumerated list of the header names (line 004). Once the enumeration of header names is returned, the `doDumpOVP()` method loops (lines 005–010) on the contents and looks up the value of each individual header by name by using the `getHeader()` method (line 008).

NOTE

If there are no headers associated with the HTTP request (a highly unlikely event), the `getHeaderNames()` method will return an empty enumerated list, not a *null*.

The doDumpParam( ) Method

The `doDumpParam()` method is similar to the `doDumpOVP()` method. In this case, however, the names and values of HTTP request parameters are dumped. Figure 4.16 shows the implementation of the `doDumpParam()` method.

Referring to Figure 4.16, we can see that the `getParameterNames()` method is used to obtain an enumerated list of parameter names (line 004). Once the enumeration of header names is returned, the `doDumpParam()` method loops (lines 005–010) on the

```
001 protected void doDumpOVP()
002 {
003   System.out.println("HTTP Header Options:");

004   Enumeration enum = m_request.getHeaderNames();

005   while (enum.hasMoreElements())
006   {
007     String strName = (String) enum.nextElement();
008     String strValue = m_request.getHeader(strName);

009     System.out.println("    " + strName + ": " + strValue);
010   }
011 }
```

Figure 4.15 The `doDumpOVP()` method.

```
001 protected void doDumpParam()
002 {
003    System.out.println("URL and/or Content Parameters:");

004    Enumeration enum = m_request.getParameterNames();

005    while (enum.hasMoreElements())
006    {
007       String strName = (String) enum.nextElement();
008       String strValue = m_request.getParameter(strName);

009       System.out.println("   " + strName + ": " + strValue);
010    }
011 }
```

Figure 4.16 The `doDumpParam()` method.

enumerated list and retrieves the value of each individual parameter by name by using the `getParameter()` method (line 008).

NOTE

If there are no parameters associated with the HTTP request, the `getParameterNames()` method will return an empty enumerated list, not a *null*.

The doHandleStats( ) Method

The `doHandleStats()` method is used to update client and session statistics. Two statistics are maintained: session hits and total hits. The session hits statistic is maintained in a session value, and the total hits statistic is maintained in a cookie stored on the requesting client. Figure 4.17 shows the implementation of the `doHandleStats()` method.

Referring to Figure 4.17, we can see that the `doHandleStats()` method begins by setting the `m_nSessionHits` variable to one (line 006) and the `nTotalHits` variable to zero (line 007). Next, the session object's `isNew()` method is called in order to see whether this request is the first request for the session at hand (line 008). If the value returned is *true*, the session is new as of this request—and there is no need to check for session values. Otherwise, if the value is *false*, the previous hit-count value for the session is retrieved via a call to `getValue()` and is incremented (lines 008–016).

Once the value of `m_nSessionHits` is established, a check is made to determine whether the value exceeds the maximum number of hits per session (line 017). If the value does exceed the maximum, the session is terminated via a call to the `invalidate()` method (line 019), and the `sendError()` method is called to send an immediate error response to the client (line 020). Otherwise, the new value of the session counter is stored in the session via a call to the `putValue()` method (line 024).

```
001 protected void doHandleStats(HttpSession session,
002   HttpServletRequest req,
003   HttpServletResponse resp)
004   throws IOException
005 {
006   int nSessionHits = 1;
007   int nTotalHits = 0;

008   if (!session.isNew())
009   {
010     Integer intSessionHits = (Integer) session.getValue("SessionHits");

011     if (intSessionHits != null)
012     {
013       nSessionHits = intSessionHits.intValue()+1;
014     }
016   }

017   if (nSessionHits > sm_nMaxSessionHits)
018   {
019     session.invalidate();
020     resp.sendError(HttpServletResponse.SC_CONFLICT);
021   }
022   else
023   {
024     session.putValue("SessionHits", new Integer(nSessionHits));

025     nTotalHits = getTotalHits(req);
026     setTotalHits(resp, ++nTotalHits);

027     System.out.println("Total Hits: " + nTotalHits);
028     System.out.println("Session Hits: " + nSessionHits);
029     System.out.println("Session Hit Max: " + sm_nMaxSessionHits);
030   }
031 }
```

Figure 4.17 The `doHandleStats()` method.

The getTotalHits( ) Method

The `getTotalHits()` method is responsible for retrieving the total hits registered by a client. This value is stored in a cookie on the client. Figure 4.18 shows the implementation of this method.

On line 003, the return value (`nTotalHits`) is initialized to zero. This value will be returned if no cookie is found. Then, an array of Cookie objects that arrived on the request is retrieved via a call to `HttpServletRequest.getCookies()`. If no cookies were present in the request, this method returns a *null*. Next, the array is searched

```
001 protected int getTotalHits(HttpServletRequest req)
002 {
003   int nTotalHits = 0;

004   Cookie cookies[] = req.getCookies();

005   if (cookies != null)
006   {
007     for (int nIndex = 0; nIndex < cookies.length; nIndex++)
008     {
009       Cookie cookie = cookies[nIndex];

010       if (cookie != null)
011       {
012         if (cookie.getName().equalsIgnoreCase("TotalHits"))
013         {
014           String strTotalHits = cookie.getValue();

015           if (strTotalHits != null)
016           {
017             try
018             {
019               nTotalHits = Integer.parseInt(strTotalHits);
020             }
021             catch (Exception e)
022             {
023               log("Invalid Hit Count ", e);
024             }
025           }
026           break;
027         }
028       }
029     }
030   }
031   return nTotalHits;
032 }
```

Figure 4.18 The `getTotalHits()` method.

(lines 007–029), looking for a cookie that has a name of "TotalHits." When the correct cookie is found, the value is converted to an integer and is assigned to `nTotalHits` (line 19).

The setTotalHits( ) Method

The `setTotalHits()` method is responsible for storing the total hits registered by a client. This value is stored in a cookie on the client. Figure 4.19 shows the implementation of this method.

```
001  protected void setTotalHits(HttpServletResponse resp, int nTotalHits)
002  {
003    Cookie cookie = new Cookie("TotalHits", String.valueOf(nTotalHits));

004    cookie.setMaxAge(60*60*24);
005    cookie.setPath("/Dump");
006    cookie.setSecure(false);

007    resp.addCookie(cookie);
008  }
```

Figure 4.19 The `setTotalHits()` method.

On line 003, a new `Cookie` class instance is created with a name of "TotalHits," and a value of the total hit counter is passed to the method. Once the cookie is instantiated, the `setMaxAge()` method is called in order to set the lifetime of the cookie to one day (line 004). In other words, 24 hours after the cookie is established (or re-established), the cookie will be released. On line 005, the path of the cookie is set for "/Dump," which is the path to the DumpServlet via a call to the `setPath()` method. Via a call to the `setSecure()` method (line 006), a secure connection is not required to receive the cookie. Finally, the `addCookie()` method is called to establish the cookie (line 007).

NOTE

When the originating browser receives the response containing the cookie, the cookie will be added (first response) or updated (subsequent responses).

The sendWMLResponse( ) Method

The last method of the DumpServlet is the only one that has anything to do with WML. While the WML document rendered here is merely a deck consisting of a single card, it does demonstrate how simple it is to render the content of an HTTP response by using the JSDK. Figure 4.20 shows the implementation of the `sendWMLResponse()` method.

Referring to Figure 4.20, the content type for the HTTP response is established on line 003 via the `setContentType()` method. Once the content type has been established, the print writer is obtained (line 005) for writing the content lines (lines 006–014). Finally, the print writer is closed (line 015) to flush the content.

Because the XML protocol requires the use of double quotes around attribute values, the XML double quotes must be escaped with a backslash. Unfortunately, this situation makes reading WML code within `println` statements a bit more difficult. Figure 4.21 shows a formatted version of the WML document rendered by the `sendWMLResponse()` method.

```
001 protected void sendWMLResponse(HttpServletResponse resp)
002  throws IOException
003 {
004  resp.setContentType("text/vnd.wap.wml");

005  PrintWriter pw = resp.getWriter();

006  pw.println("<?xml version=\"1.0\"?>");
007  pw.print("<!DOCTYPE wml PUBLIC \"");
008  pw.print("-//WAPFORUM//DTD WML 1.1//EN\" ");
009  pw.println("\"http://www.wapforum.org/DTD/wml_1.1.xml\">");
010  pw.println("<wml>");
011  pw.println("<card>");
012  pw.println("<p>See output stream for http header info</p>");
013  pw.println("</card>");
014  pw.println("</wml>");
015  pw.close();
016 }
```

Figure 4.20 The `sendWMLResponse()` method.

```
<?xml version="1.0"?>
<!DOCTYPE wml PUBLIC "-//WAPFORUM//DTD WML 1.1//EN"
 "http://www.wapforum.org/DTD/wml_1.1.xml">

<wml>
 <card>
  <p>See output stream for http header info</p>
 </card>
</wml>
```

Figure 4.21 WML content rendered by the `sendWMLResponse()` method.

Executing the Servlet

In Chapter 3, we saw the DumpServlet in action. We did not see the statistics portion, however. Therefore, in this chapter, we will examine that feature. Specify the following URL in your WAP SDK browser:

```
http://localhost/Dump?Stats=Yes
```

For our purposes, we will load this URL and reload it three additional times. This action will cause the session counter to increase from one to three. On the final reload, the error code will be returned due to the session count going beyond its maximum value. Figure 4.22a shows a progression of screen shots associated with the example. Panel

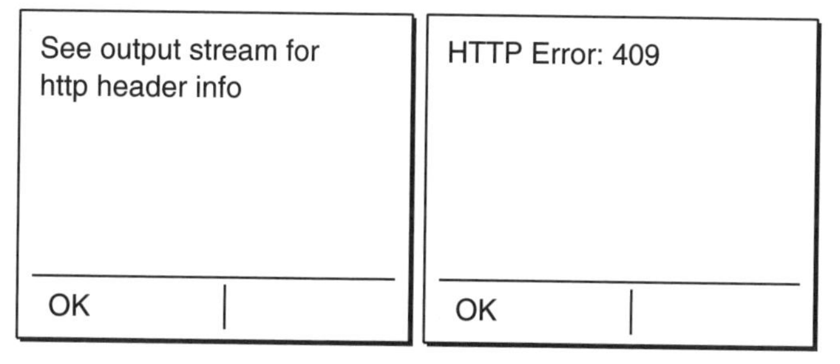

Figure 4.22a Screen shots from the DumpServlet.

```
Total Hits: 1
Session Hits: 1
Session Hit Max: 3
Total Hits: 2
Session Hits: 2
Session Hit Max: 3
Total Hits: 3
Session Hits: 3
Session Hit Max: 3
```

Figure 4.22b Output from the DumpServlet.

one shows the view of the first three loads, and panel two shows the screen of the last load. Figure 4.22b shows the statistics output written to `System.out`.

Conclusion

In this chapter, we looked at the JSDK classes and interfaces that relate best to the task of writing WAP Servlets: `HttpServlet`, `HttpSession`, `HttpServletRequest`, `HttpSerlvetResponse`, and `Cookie`. We also examined the methods from these classes and interfaces that are essential for writing WAP Servlets. Finally, to reinforce our understanding of the methods, we looked at the DumpServlet as a simple JSDK example.

In Chapter 5, we will combine what we learned about the JSDK in this chapter and what we learned about WML in Chapter 3 and develop a real-world WAP Servlet.

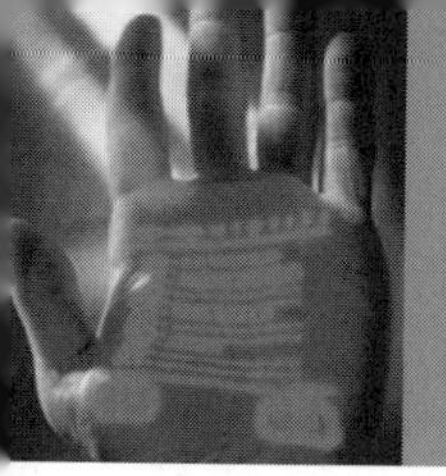

CHAPTER 5

Wireless Markup Language (WML) Homepage Example

You might classify *Hypertext Transport Protocol* (HTTP)-based Web content by placing it into two general categories: static content and dynamic content. People usually think of static content in terms of static data files (containing markup) that a standard HTTP server serves. Dynamic content is typically associated with Common Gateway Interface (CGI) scripts and servlets. This chapter explores both of these areas by examining the subject of homepages.

Introduction

In Chapter 4, "*The Java Servlet Development Kit* (JSDK)," we learned about developing with JSDK. The `DumpServlet` example from that chapter provided us with a glimpse into Wireless Access Protocol (WAP) servlet development by rendering a small portion of Wireless Markup Language (WML) content. In this chapter, we take the next step by examining a real-world servlet example.

Much of the content of a homepage (whether it is an individual or corporation homepage) is static in nature. Even if the content changes over time, Webmasters find it easy to use popular Web packages in order to maintain static markup language files that are linked and are served as *Multi-Purpose Internet Mail Extensions* (MIME) content by an HTTP server. When forms are embedded within a site's pages, a CGI script or a servlet must exist in order to process the collected data—but it is possible, and often the case, that the content itself is served from files rather than being rendered on the fly.

As enterprise servers become more popular, the notion of rendering such content will become more popular. On that note, we will explore two scenarios in this chapter. From an end-user's perspective, the homepage that we generate with these examples will appear exactly the same as a regular homepage. The development techniques will be different, however. In the first example, we will briefly look at serving WML content from files. In the second example, we will look (in a bit more detail) at how you can render the same basic information dynamically.

While some readers might find a discussion of homepage development basic in nature, the examples (especially the second example) build upon the information from prior chapters and provide a foundation for the chapters to come.

Simply Serving Content

As stated in the previous section, serving content rather than rendering it is not only a totally acceptable method of producing a Web presence, but in many cases, it is the preferred technique. *Hypertext Markup Language* (HTML) is much like chess: learning the basics is relatively easy, but it takes a lifetime to learn and master the subtleties. Obviously, there are many HTML editors available on the market today. These editors make the task of remembering and using HTML elements and attributes manageable. Additionally, many popular office automation tools have the capability to output HTML content. This capability usually means that if you know how to use a word processor, for example, you can generate HTML files without knowing much about HTML. Unfortunately, due to the fact that most of these programs do not have a capability to generate WML, the same statement does not hold true for WAP sites.

Consider Webmasters who have an excellent HTML presence on the Web. They would like to extend their presence by rendering similar content for WAP browsers via WML. If these Webmasters understand HTML and prefer to directly edit the source of their content, the notion of simply serving hand-crafted WML content might seem entirely reasonable. After all, WML is just another markup language. This section explores this technique.

You will recall from previous chapters that the companion CD-ROM included with this book has a `/code` directory that contains source code for the examples provided. The code directory contains a WML document that serves as the WML index to the book: `/code/index.wml`. Because the servlet engine on the CD-ROM considers the `/code` directory to be its public document directory, you can access the `/code/index.wml` document from your WAP simulator of choice by specifying the URL `http://localhost/index.wml`.

In general, Figure 5.1 shows the WML deck contained within this file. Specifically, this deck consists of two cards: card1 and card2. The function of card1 is to show the title of the book, establish a *prev* soft key, and associate the *accept* soft key with card2. The function of card2 is to show a list of links to other WML documents that are associated with the book's chapters.

TIP

As stated previously, card2 (from the example shown in Figure 5.1) provides a list of related links. While it is possible to use a `<select>` element to achieve the same results on some WAP browser implementations, experience has driven me to use

```
<?xml version="1.0"?>
<!DOCTYPE wml PUBLIC "-//WAPFORUM//DTD WML 1.1//EN"
"http://www.wapforum.org/DTD/wml_1.1.xml">
<wml>
  <template>
    <do type="prev" label="Back">
      <prev/>
    </do>
  </template>
  <card title="Book" id="card1">
    <do type="accept" label="Links">
      <go href="#card2"/>
    </do>
    <do type="prev" label="Back">
      <prev/>
    </do>
    <p align="center">
      <b>Book</b>
    </p>
    <p>
      WAP Servlets Developers Guide by John Cook
    </p>
  </card>
  <card id="card2">
    <p>
      Chapters:<br/>
      <a href="/book/chap01/ch01.wml">Chapter 1</a>
      <br/>
      <a href="/book/chap02/ch02.wml">Chapter 2</a>
      <br/>
      <a href="/book/chap03/ch03.wml">Chapter 3</a>
      <br/>
      <a href="/book/chap04/ch04.wml">Chapter 4</a>
      <br/>
      <a href="/book/chap05/ch05.wml">Chapter 5</a>
      <br/>
      <a href="/book/chap06/ch06.wml">Chapter 6</a>
      <br/>
      <a href="/book/chap07/ch07.wml">Chapter 7</a>
      <br/>
      <a href="/book/chap08/ch08.wml">Chapter 8</a>
      <br/>
      <a href="/book/chap09/ch09.wml">Chapter 9</a>
      <br/>
      <a href="/book/chap10/ch10.wml">Chapter 10</a>
      <br/>
```

(continues)

Figure 5.1 Source for `/code/index.wml`.

```
        <a href="/book/appa.wml">Appendix A</a>
        <br/>
        <a href="/book/appb.wml">Appendix B</a>
        <br/>
        <a href="/book/appc.wml">Appendix C</a>
        <br/>
        <a href="/book/appd.wml">Appendix D</a>
      </p>
    </card>
  </wml>
```

Figure 5.1 Source for `/code/index.wml` (Continued).

hypertext links in my WML work. Using hypertext links typically provides the most consistent look and feel across various browser implementations.

NOTE

The example in Figure 5.1 (and in fact, most of the decks in this chapter) uses the `<template>` element to implement a common `<do>` element of type prev. Some browser implementations do not provide a default back function associated with the *accept* key. Therefore, you should always define an explicit prev type `<do>` element. Otherwise, users might find themselves in a dead end.

Each of the hypertext links provided in card2 refer to documents within each of the respective chapter directories. Each of the links are specified as a relative URL (relative to the server that served the index file). You can think of these files as index files for a particular chapter. These chapter index files come in two flavors: *description only* and *description with links*. I show and describe both of these types in the following paragraphs.

The WML document in Figure 5.2 shows the *description-only* form of a chapter index file. Specifically, the example shows the index file for Chapter 1: `/book/chap01/ch01.wml`. This WML deck consists of one card: card1. The function of card1 is to display a title (such as Chapter 1) and a description of the chapter. Once again, we establish a *prev* soft key in order to navigate backwards.

Figure 5.3 shows a progression of screen shots. Specifically, the first panel shows the title page from the index file (Figure 5.1). Once the user presses the *Links* (*accept*) soft key, the second panel displays the links to the chapters. Once the Chapter 1 link is selected via the *Link (accept)* soft key, the third panel shows the title and description of Chapter 1 (Figure 5.2).

The WML document in Figure 5.4 shows the description with links form of a chapter index file. Specifically, the example shows the index file for Chapter 5: `/book/chap05/ch05.wml`. This WML deck consists of two cards: card1 and card2. The function of card1 is to display a title (such as Chapter 5) and a description of the chapter. As

```
<?xml version="1.0"?>
<!DOCTYPE wml PUBLIC "-//WAPFORUM//DTD WML 1.1//EN"
"http://www.wapforum.org/DTD/wml_1.1.xml">
<wml>
  <card title="Chapter 1" id="card1">
    <do type="prev" label="Back">
      <prev/>
    </do>
    <p align="center">
      <b>Chapter 1</b>
    </p>
    <p>
      This chapter is the introduction to the book and its contents.
    </p>
  </card>
</wml>
```

Figure 5.2 Source for `/code/book/chap01/ch01.wml`.

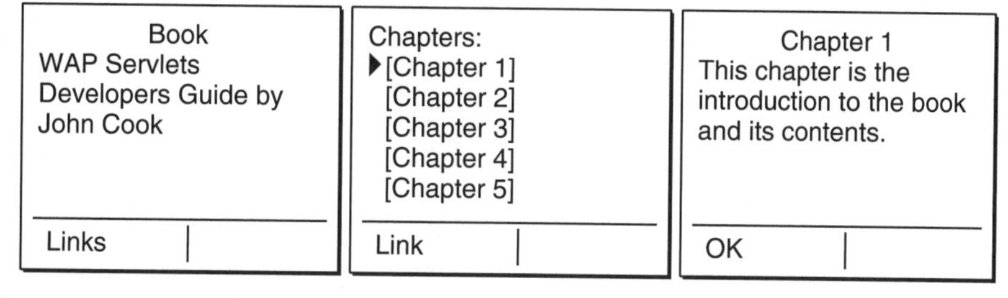

Figure 5.3 Screen shots associated with the no link chapter index.

before, we establish the *prev* soft key in order to navigate backwards. Furthermore, the *accept* soft key is associated with card2. Because this chapter contains two examples, card2 establishes hypertext links to two additional WML documents that provide descriptions of the examples.

NOTE

Because the two WML documents associated with this chapter's examples reside within the same directory as the chapter's index file, card2 utilizes relative URLs in its hypertext links (relative to the chapter's index file).

The WML document shown in Figure 5.5 is the one associated with the first example. Specifically, this example shows a deck consisting of a single card: card1. As usual, a *prev* soft key is established. This card also presents a title and a description. Therefore, the example card looks similar in structure to the *description-only* form of the chapter index files; in fact, they are the same structurally.

```
<?xml version="1.0"?>
<!DOCTYPE wml PUBLIC "-//WAPFORUM//DTD WML 1.1//EN"
"http://www.wapforum.org/DTD/wml_1.1.xml">
<wml>
  <template>
    <do type="prev" label="Back">
      <prev/>
    </do>
  </template>
  <card title="Chapter 5" id="card1">
    <do type="accept" label="Links">
      <go href="#card2"/>
    </do>
    <p align="center">
      <b>Chapter 5</b>
    </p>
    <p>
      This chapter is focused on serving simple static wml content.
      Two examples are provided. Follow links for more information
      about the examples.
    </p>
  </card>
  <card id="card2">
    <p>
      Examples:<br/>
      <a href="ch05ex01.wml">Example 1</a>
      <br/>
      <a href="ch05ex02.wml">Example 2</a>
      <br/>
    </p>
  </card>
</wml>
```

Figure 5.4 Source for `/code/book/chap05/ch05.wml`.

Figure 5.6 shows a progression of screen shots. Specifically, the first panel shows the description of Chapter 5. Once the user presses the *Links (accept)* soft key, the second panel is shown (showing the links to the two examples). Once the Example 1 link is selected via the *Link (accept)* soft key, the third panel is shown (showing the description of the example).

Relatively speaking, the WML documents served in this example are simple. Using the basic techniques explored here, however, you could implement a WML Web site that consists of many levels. By using additional techniques, you could implement a more complex means of navigating between pages than the tree orientation used here. For many Webmasters seeking to get some WML content out quickly, however, this methodology might provide a good boilerplate.

```
<?xml version="1.0"?>
<!DOCTYPE wml PUBLIC "-//WAPFORUM//DTD WML 1.1//EN"
"http://www.wapforum.org/DTD/wml_1.1.xml">
<wml>
  <card title="Example 1" id="card1">
    <do type="prev" label="Back">
      <prev/>
    </do>
    <p align="center">
      <b>Example 1</b>
    </p>
    <p>
      This example shows how the HTTP server included with the book
      can be used to server static WML documents.
    </p>
  </card>
</wml>
```

Figure 5.5 Source for `/code/book/chap05/ch05ex01.wml`.

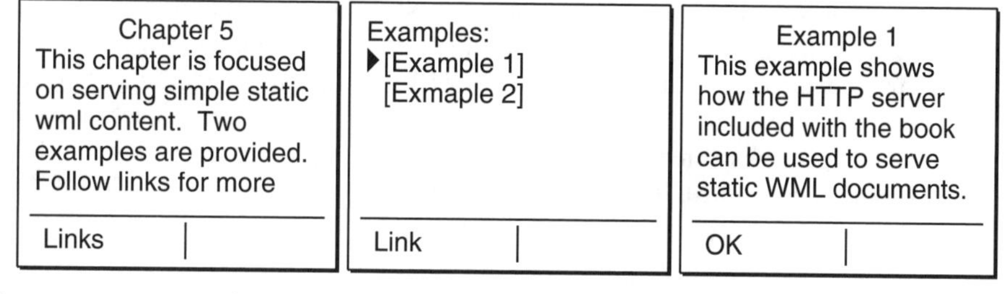

Figure 5.6 Screen shots associated with a linked chapter index.

Execution

In order to see for yourself the various WML decks being served by an HTTP servlet, you can use the servlet engine from the accompanying CD-ROM. To execute the server, perform the following procedures:

- Change directories to the `/code` directory on the CD-ROM or to a directory that you copied the contents to for experimentation purposes.
- Make sure that the directory "." is in your CLASSPATH environment variable.
- Make sure that the `java.exe` executable from the JDK is in your execution search path.
- Enter `java com.n1guz.ServletEngine.ServletEngine` at the command line.

TIP

Because the `PageView` servlet is pure Java, you should be able to run it from any HTTP server that supports JSDK. In order for the server to find the WML documents, however, you will need to configure its public directory to the `/code` directory. Alternatively, you can move the contents of the `/code` directory.

Once the server is online, you can start your favorite WAP browser simulator and execute the following URL to see the served content: http:// index.wml.

Rendering Content

In the previous section, we looked at serving WML content that resides in server documents. In this section, we will construct a servlet that is capable of rendering the same content from a description contained in a data file. One primary advantage of rendering versus serving is that you can maintain the data for the entire site in a single data file.

Requirements

Because we will be writing general-purpose software in the form of a servlet, we should start with a list of requirements. In the previous section, we saw that a reasonable, static Web site can be implemented in WML by using two basic kinds of decks: single card and two card. The single-card form was used for decks that described something. The two-card decks consisted of a description on the first card and a set of hypertext links on the second card. In both cases, all cards have a title and a header. The following list describes the requirements for the servlet:

- The primary function of the servlet is to render pages.
- Pages must have a heading that is bold and centered at the top.
- Pages can have optional text.
- Pages can have optional links.
- The page link can refer to other pages or URLs.
- All pages that the servlet serves must be described in a single data file.

Design

The next task is to design our servlet. By examining these requirements, we can establish the following design elements:

- A data file `servlet.dat` contains data used to construct all pages associated with the servlet. This file will reside in the package directory of the servlet.
- Attributes of the pages contained in the data files will be specified as name-value pairs with the format `name:value`.

- Each time the page attribute is encountered, a new page will be instantiated. The value of the page attribute will be the internal name of the page (used for linking). The first page encountered is the default page.
- If a ViewPage parameter is specified on the URL used to invoke the servlet or within the content of a postrequest, the value of that parameter is matched against a list of internal page names maintained by the servlet. If a match is found, that page is rendered as the content of the HTTP response. If a match is not found, the HTTP error 404 (resource not found) is returned. If no ViewPage parameter is specified, the default page is rendered.
- Pages are implemented as WML decks consisting of two cards. The first card contains the heading and the text (if specified). The second card contains links associated with the page (if specified). If no links are specified for a particular page, neither the second card nor its link from the first card will be rendered.
- Within the servlet data file, the title attribute is used to establish the title of the page at hand. The page title is used as the value of the title attribute of the first <card> element when the page is rendered.
- Within the servlet data file, the heading attribute is used to establish the heading of the page at hand. The page heading is displayed as bold, centered text at the top of the first <card> when the page is rendered. If a heading attribute is not specified for a given page, the default value is used. The default value is the value of the title attribute.
- Within the servlet data file, the text attribute is used to add text to the page. If more than one text attribute is specified for a given page, all values are concatenated to form a single text value for the page.
- Within the servlet data file, the LinkHeading attribute is used to establish the heading of the second <card> element for the page. The link heading is displayed within the card as left-aligned, bold text. If the LinkHeading attribute is not specified for a given page, the default value Links is used.
- Within the servlet data file, the Link attribute is used to add a link to the list of links associated with a page. The value of the Link attribute is a name-value pair consisting of a label and a link with the format *label:link*. Therefore, the format of the Link attribute is as follows: *Link:label:link*. If a link attribute is specified, it must have both a label and a link. If the link value starts with a pound sign (#), it is assumed to be an internal page name; otherwise, the link is assumed to be an absolute URL.
- The servlet data file can contain empty lines in order to improve readability.

The Servlet Data File

Based on the servlet design defined earlier, Figure 5.7 shows the servlet.dat file that mimics the behavior of the served documents from the first section. For example, the lines in the figure that follow Page:index represent the document shown in Figure 5.1. Not only is the format of each page simpler than its WML counterpart, but the

```
Page:index
Title:Book
Heading:WAP Servlets Developers Guide by John Cook
LinkHeading:Chapters
Link:Chapter 1:#ch01
Link:Chapter 2:#ch02
Link:Chapter 3:#ch03
Link:Chapter 4:#ch04
Link:Chapter 5:#ch05
Link:Chapter 6:#ch06
Link:Chapter 7:#ch07
Link:Chapter 8:#ch08
Link:Chapter 9:#ch09
Link:Appendix A:#apa
Link:Appendix B:#apb
Link:Appendix C:#apc
Link:Appendix D:#apd

Page:ch01
Title:Chapter 1
Text:This chapter is the introduction to the book and its contents.

Page:ch02
Title:Chapter 2
Text:This chapter provides background information about wired and
Text:wireless networking.

Page:ch03
Title:Chapter 3
Text:This chapter discusses the basics of XML and provides a tutorial
on WML
Text:development. The WML tutorial describes the various elements
Text:and attributes of WML as well as providing examples.

Page:ch04
Title:Chapter 4
Text:This chapter discusses the JSDK and the basics of servlet
development.
Text:It also shows an example JSDK servlet: the DumpServlet. The
Text:DumpServlet is used in Chapter 3 to dump the header information
Text:in some examples.

Page:ch05
Title:Chapter 5
Text:This chapter is focused on serving simple static wml content.
Text:Two examples are provided. Follow links for more information
Text:about the examples.
```

Figure 5.7 Source for /code/book/chap05/PageViewer/servlet.dat.

```
LinkHeading:Examples
Link:Example 1:#ch05ex01
Link:Example 2:#ch05ex02

Page:ch05ex01
Title:Example 1
Text:This example shows how the HTTP server included with the book
Text:can be used to server static WML documents.

Page:ch05ex02
Title:Example 2
Text:This example is similar in function to example 1; however, the
Text:static content is rendered dynamically by a servlet based on
Text:the content of a single text file.

Page:ch06
Title:Chapter 6
Text:This chapter presents a real world application: "Grocery."
Text:Follow links for more information about the example.
LinkHeading:Example
Link:ReadMe:#ch06ex01
Link:Grocery:/Grocery

Page:ch06ex01
Title:Grocery Example
Text:This example allows users to construct a shopping list
Text:that they can later access from the grocery store.

Page:ch07
Title:Chapter 7
Text:This chapter discusses the push technology associated with
Text:WAP.

Page:ch08
Title:Chapter 8
Text:This chapter examines WML Script.
Text:Follow links for more information about the example.
LinkHeading:Example
Link:ReadMe:#ch08ex01
Link:Blackjack:/Blackjack

Page:ch08ex01
Title:Blackjack Example
Text:Using this example, users can play Blackjack via a
Text:WML Script.
```

(continues)

Figure 5.7 Continued.

```
Page:ch09
Title:Chapter 9
Text:This chapter concludes the book by making recommendations
Text:on the next step that developers may take with their
Text:WAP development.

Page:apa
Title:Appendix A
Text:Provides a reference to the WAP forums WML 1.2 specification.

Page:apb
Title:Appendix B
Text:Provides a reference to Sun's JSDK 2.1.1

Page:apc
Title:Appendix C
Text:Provides a source listing of the servlet engine.

Page:apd
Title:Appendix D
Text:Describes the contents of the companion CDROM.
```

Figure 5.7 Source for `/code/book/chap05/PageViewer/servlet.dat` (Continued).

capability to specify all data within a single file is much easier than dealing with a WML document set that spans a file system.

The Implementation

With the requirements and the design behind us, we are now ready to implement the servlet. Having a format for the servlet data file and a sample data file is also useful. The implementation of the servlet is realized in three classes: `Page`, `PageLink`, and `PageViewer`. All three of these classes are defined within the `book.chap05.PageViewer` package; thus, they are located in the directory `/code/book/chap05/PageViewer` on the companion CD-ROM. The following sections examine the function of these classes in more detail.

The Page Class

The `Page` class represents a page to be rendered by the servlet. When the servlet initializes, the servlet data file is read. As each page is constructed from the data file, it is stored in an instance of the `Page` class. Figure 5.8 shows the definition of the `Page` class.

As shown in the figure, the implementation of the `Page` class is simple. In fact, the class is merely a data structure with members associated with the attributes of a page and

```
package book.chap05.PageViewer;

import java.util.*;

public class Page
{
    protected String m_strName = null;
    protected String m_strTitle = null;
    protected String m_strHeading = null;
    protected Vector m_vTextLines = null;
    protected String m_strLinkHeading = "Links";
    protected Vector m_vLinks = null;

    public Page(String strName)
    {
        m_strName = strName;
    }

    public void setTitle(String strTitle)
    {
        m_strTitle = strTitle;

        if (m_strHeading == null)
        {
            m_strHeading = m_strTitle;
        }
    }

    public String getTitle()
    {
        return m_strTitle;
    }

    public void setHeading(String strHeading)
    {
        m_strHeading = strHeading;
    }

    public String getHeading()
    {
        return m_strHeading;
    }

    public void addTextLine(String strTextLine)
    {
        if (m_vTextLines == null)
```

(continues)

Figure 5.8 The `Page` class.

```
        {
            m_vTextLines = new Vector();
        }

        m_vTextLines.addElement(strTextLine);
    }

    public Enumeration getTextLines()
    {
        Enumeration retval = null;

        if (m_vTextLines != null)
        {
            retval = m_vTextLines.elements();
        }

        return retval;
    }

    public void setLinkHeading(String strLinkHeading)
    {
        m_strLinkHeading = strLinkHeading;
    }

    public String getLinkHeading()
    {
        return m_strLinkHeading;
    }

    public void addLink(PageLink pageLink)
    {
        if (m_vLinks == null)
        {
            m_vLinks = new Vector();
        }

        m_vLinks.addElement(pageLink);
    }

    public Enumeration getLinks()
    {
        Enumeration retval = null;

        if (m_vLinks != null)
        {
            retval = m_vLinks.elements();
        }
```

Figure 5.8 The `Page` class (Continued).

```
            return retval;
        }
    }
```

Figure 5.8 Continued.

convenience methods for setting and getting those attributes. Due to the simple nature of this class, a line-by-line explanation is not required.

The PageLink Class

The `PageLink` class represents a link from one page to either another page or to a URL. As each page is constructed from the servlet data file, its links are stored in instances of the `PageLink` class. Figure 5.9 shows the implementation of the `PageLink` class. Once again, the `PageLink` class is straightforward, so a detailed discussion is not necessary.

The PageViewer Class

In previous sections, we looked quickly at the two support classes of this servlet: `Page` and `PageLink`. Unlike these undemanding utility classes, the `PageViewer` class is a bit more complex and deserves a detailed examination. After all, this class *is* the servlet. Figure 5.10 shows the definition of the class.

As shown in the figure, the `PageViewer` class consists of several data members and methods. The following sections will explore the contents of the class in greater detail.

Imports

The import statements of the `PageViewer` class are shown in Figure 5.11. The `java.io` package is imported primarily to support the effort of reading the `servlet.dat` file (which contains the page definitions). The `java.util` package is imported primarily for `Hashtable`, `Vector`, and `Enumeration`. Both of the JSDK packages are supported, because `PageViewer` is a servlet.

Data Members

Two data members are also defined: `m_firstPage` and `m_hashPages` (see Figure 5.10). These values are worth putting into servlet class members, because they will be frequently referenced and they are non-volatile with respect to individual requests. If these values were volatile (in other words, requests competing over their value), they would need to be stored in automatic variables or as `HttpSession` object values.

```
package book.chap05.PageViewer;

public class PageLink
{
    protected String m_strLabel = null;
    protected String m_strLink = null;

    public PageLink (String strLabel, String strLink)
    {
        if ((strLabel == null) || (strLabel.length() == 0))
        {
            throw new IllegalArgumentException();
        }

        if ((strLink == null) || (strLink.length() == 0))
        {
            throw new IllegalArgumentException();
        }

        m_strLabel = strLabel;
        m_strLink = strLink;
    }

    public String getLabel()
    {
        return m_strLabel;
    }

    public String getLink()
    {
        return m_strLink;
    }
}
```

Figure 5.9 The `PageLink` class.

TIP

Remember that the data members of the servlet are shared by all requests made to the servlet. While each request executes within its own thread, they all use the same instance of the object. Because `m_hashPages` and `m_dataFile` are servlet data rather than user (or session) data, I have placed them at the class level. In a standard application, you might be tempted to make these members static. I intentionally did not make them static, however. If they were static, we could not run two independent instances of the servlet in order to serve different Web sites. The second site would wipe out the `m_hashPages` member when it received its first request (or otherwise initialized).

```
public class PageViewer extends HttpServlet
{
    protected Page m_firstPage = null;

    protected Hashtable m_hashPages = new Hashtable();

    public void init(ServletConfig config)
        throws ServletException;

    protected void loadDataFile(String strDataFile);

    protected void loadDataFile(File file);

    protected void loadDataFile(BufferedReader br)
        throws IOException;

    protected void processLink(Page page,
        String strLinkData, int nLine);

    public String getServletInfo();

    public void doGet(HttpServletRequest req, HttpServletResponse resp)
        throws ServletException, IOException;

    public void printPage(Page page, PrintWriter pw);

    protected void printLinkCard(PrintWriter pw,
        String strHeading, Enumeration enumLinks);
}
```

Figure 5.10 Definition of the `PageViewer` class.

```
import java.io.*;
import java.util.*;
import javax.servlet.*;
import javax.servlet.http.*;
```

Figure 5.11 `PageViewer` class imports.

WARNING

This point is so important that it is worth saying again. Unless there is a compelling reason for a servlet to have static members, you should not use them. You prevent the servlet from working correctly when the same servlet is associated with different databases.

```
001 public void init(ServletConfig config) throws ServletException
002 {
003    super.init(config);
004    String dataFile = config.getInitParameter("datafile");
005    loadDataFile(dataFile);
006 }
```

Figure 5.12 The `init()` method.

```
01 protected void loadDataFile(String strDataFile)
02 {
03     File file = new File(strDataFile);

04     String strAbsolutePath = file.getAbsolutePath();

05     if (file.exists())
06     {
07         log("Loading data file '" + strAbsolutePath + ".'");
08         loadDataFile(file);
09     }
10     else
11     {
12         log("Can't find data file '" + strAbsolutePath + ".'");
13     }
14 }
```

Figure 5.13 The `loadDataFile(String)` method.

The init( ) Method

The implementation of the `init()` method appears in Figure 5.12. As discussed previously, the `init()` function initializes the static members of the servlet. Unlike the `DumpServlet` method (from the previous chapter), the `PageViewer` servlet has meaningful initialization work to accomplish.

On line 3, the `init()` method of the super class is called so that it can initialize itself based on the `ServletConfig` object passed into the servlet by the HTTP server. Next, the `ServletConfig.getInitParameter()` method is called in order to get the path to the datafile (line 004). Next, the `loadDataFile()` method is called to load the `servlet.dat` file into the `m_hashPages` member. Finally, the data file is loaded via `loadDataFile()`.

The loadDataFile(String) Methods

The `loadDataFile()` method is so busy that it is broken down over three methods. The first form of the method, `loadDataFile(String)`, is shown in Figure 5.13. This form of the method takes the relative name of the servlet.dat file containing the page definitions relative to the package root /code. The primary function of this method is to take the filename passed as a parameter, instantiating a File object for that filename and verifying that the file exists.

On line 03, a File object is constructed for the servlet.dat file. If the file exists (line 05), a message is written to the HTTP server log indicating that the data file is being loaded (line 07). Next, the second form of the loadDataFile(File) method is called on that file object (line 08). If, however, the file does not exist, a message is written to the HTTP server log indicating that the file cannot be found (line 12).

The loadDataFile(File) Method

The loadDataFile(File) method picks up the work of loading the servlet.dat file. The implementation of this method is shown in Figure 5.14. The primary function

```
01 protected void loadDataFile(File file)
02 {
03     BufferedReader br = null;

04     try
05     {
06         FileInputStream fis = new FileInputStream(file);
07         InputStreamReader isr = new InputStreamReader(fis);
08         br = new BufferedReader(isr);
09     }
10     catch (IOException e)
11     {
12         br = null;
13     }

14     if (br != null)
15     {
16         try
17         {
18             loadDataFile(br);
19         }
20         catch (IOException e)
21         {
22             // End of file.
23         }
24     }
25     else
26     {
27         log("Unable to read '" + file.getAbsolutePath() + ".'");
28     }
25 }
```

Figure 5.14 The loadDataFile(File) method.

of this method is to take the `File` object passed into the method, open an input stream, and wrap it with a `BufferedReader`.

The first order of business in this method is to wrap the `File` object passed as a parameter with a `FileInputStream` object (line 06), an `InputStreamReader` object (line 07), and a `BufferedReader` object (line 08). If all goes well and the buffered reader is successfully created, the next step is to call the `loadDataFile(BufferedReader)` method (line 18). If for some reason the `BufferedReader` is not successfully created, a message is written to the HTTP server log (line 27).

NOTE

Because the `server.dat` file is an ASCII file, we want to use a Reader to access its data, due to the fact that `loadDataFile(BufferedReader)` reads lines from the file. For efficient input, we want to use a buffered reader. The `BufferedReader` class also provides a `readLine()` function to read a line into a string.

The loadDataFile(BufferedReader) Method

Now that we have opened a file and wrapped it with a `BufferedReader`, we are ready to start reading the `servlet.dat` file and construct `Page` objects based on its content. This function is the primary function of this method. The implementation of the method is shown in Figure 5.15.

Upon entering the `loadDataFile(BufferedReader)` method, the `page` variable is established for holding the current page (the page that is in the process of being constructed from data as it is read from the `servlet.dat` file), represented in line 4. Also, the `nLine` variable is established and initialized to zero. This variable provides the current line count so that the line number can be included with errors reported to the log. Once the initial variables are established, the balance of time spent in this method involves looping through the lines in the file (lines 06–59).

The main loop of the method continues as long as there are lines to be read from the data file (line 06). The first task within the loop is to read a line from the file and trim any leading or trailing white space (line 08). Next, the `nLine` variable is incremented so that the line number reported on logged errors will match the line just read (line 09). At that point, the length of the line is obtained (line 10), and the loop is cycled if the line is empty (lines 11–14).

NOTE

Recall from the requirements that blank lines are allowed for readability.

Using the length of the line and the location of the colon delimiting the tag from the data (line 15), the structure of the line is analyzed for errors. The system checks for missing colons (lines 16–21) and for zero-length data portions (lines 22–27). In either case, if an error is discovered, the error is logged to the HTTP server log, and cycling the loop skips the line. If a structural error is not found on the line, the tag portion (line 28) and the data portion (line 29) are extracted from the line.

At this point, the value of the tag is tested for the keywords established in the servlet design: `Page` (line 30), `Title` (line 39), `Heading` (line 43), `Text` (line 47), `Link` (line

```
01 protected void loadDataFile(BufferedReader br)
02     throws IOException
03 {
04     Page page = null;

05     int nLine = 0;

06     while(br.ready())
07     {
08         String strLine = br.readLine().trim();
09         nLine++;

10         int nLength = strLine.length();

11         if (nLength == 0)
12         {
13             continue;
14         }

15         int nPos = strLine.indexOf(':');

16         if (nPos == -1)
17         {
18             log("Line " + nLine +
19                 ": invalid name:value format.");
20             continue;
21         }

22         if (nPos == (nLength-1))
23         {
24             log("Line " + nLine +
25                 ": name:value pair has zero length value.");
26             continue;
27         }

28         String strTag = strLine.substring(0,nPos).trim();
29         String strData = strLine.substring(nPos+1).trim();

30         if (strTag.equalsIgnoreCase("Page"))
31         {
32             page = new Page(strData);
33             m_hashPages.put(strData, page);

34             if (m_firstPage == null)
35             {
36                 m_firstPage = page;
```

(continues)

Figure 5.15 The `loadDataFile(BufferedReader)` method.

```
37                  }
38              }
39              else if (strTag.equalsIgnoreCase("Title"))
40              {
41                  page.setTitle(strData);
42              }
43              else if (strTag.equalsIgnoreCase("Heading"))
44              {
45                  page.setHeading(strData);
46              }
47              else if (strTag.equalsIgnoreCase("Text"))
48              {
49                  page.addTextLine(strData);
50              }
51              else if (strTag.equalsIgnoreCase("Link"))
52              {
53                  processLink(page, strData, nLine);
54              }
55              else if (strTag.equalsIgnoreCase("LinkHeading"))
56              {
57                  page.setLinkHeading(strData);
58              }
59          }
60  }
```

Figure 5.15 Continued.

51), and `LinkHeading` (line 55). As defined in the design, the detection of the `Page` keyword indicates the start of a new page. Creating a new `Page` object consists of instantiating a new Page object and setting its internal name (line 32). Once the new `Page` object has been created, it is stored in the `m_hashPages` hash table by its internal name (line 33). If the `m_firstPage` member is still a `null` (indicating that this page is the first page within the data file), the value of the `m_firstPage` member is set to the current page (lines 34–37).

When the `Link` keyword is detected (line 51), a `LinkData` object is instantiated and associated with the current page via a call to the `processLink()` method (line 53). Otherwise, as the other keywords are detected, the appropriate convenience method of the `Page` class is invoked in order to establish the value associated with the current data file line.

The processLink( ) Method

During the reading of the servlet data file, most tags were processed within the `loadDataFile(BufferedReader)` method. When a `Link` tag is encountered, the `processLink()` method is called. The `processLink()` method is used to process the value of the `Link` tag from the servlet data file, due to the fact that the `Link` tag is more complex than the other tags. The implementation of this method is shown in Fig-

```
001 protected void processLink(Page page,
002     String strLinkData, int nLine)
003 {
004     int nLength = strLinkData.length();
005     int nPos = strLinkData.indexOf(':');

006     if ((nPos == -1) || (nPos == (nLength-1)))
007     {
008         log("Line " + nLine + ": invalid link entry.");
009         log("Correct format is 'Link:Label:(#PageName|URL)'");
010     }
011     else
012     {
013         String strLabel = strLinkData.substring(0,nPos).trim();
014         String strLink = strLinkData.substring(nPos+1).trim();

015         PageLink pageLink = new PageLink(strLabel, strLink);
016         page.addLink(pageLink);
017     }
018 }
```

Figure 5.16 The `processLink()` method.

ure 5.16. The primary function of this method is to separate the label and link portions of a `Link` tag's value.

Based on the servlet design, the value of the `Link` tag (from the servlet data file) consists of two parts separated by a colon: the label and the link. Upon entry into the `processLink()` method, the length of the value (line 4) and the position of the colon (line 5) are calculated. Next, these values are analyzed for correctness (line 6). If the values are incorrect (indicating either a missing colon or a zero-length link), an error message is written to the log (lines 8–9). Otherwise, the label (line 13) and link (line 14) portions of the value are parsed, and these values are used to construct a new `PageLink` object (line 15). Upon successfully constructing the new `PageLink` object, the object is added to the current `Page` object via the `addLink()` method.

The getServletInfo( ) Method

As discussed earlier, the `getServletInfo()` method is one of the standard methods from the `HttpServlet` class that enables the host HTTP server to collect information about the servlet. The implementation of the `getServletInfo()` method is shown in Figure 5.17.

The doGet( ) Method

The `doGet()` method is the code that services HTTP get requests received by the HTTP servlet targeting the servlet. The implementation of the method is shown in

```
001 public String getServletInfo()
002 {
003     return "PageViewer by John Cook";
004 }
```

Figure 5.17 The `getServletInfo()` method.

```
001 public void doGet(HttpServletRequest req,
        HttpServletResponse resp)
002     throws ServletException, IOException
003 {

004     String strPageName = req.getParameter("ViewPage");

005     Page page = null;

006     if (strPageName != null)
007     {
008         page = (Page) m_hashPages.get(strPageName);
009     }
010     else
011     {
012         page = m_firstPage;
013     }

014     if (page != null)
015     {
016         resp.setContentType("text/vnd.wap.wml");
017         printPage(page, resp.getWriter());
018     }
019     else
020     {
021         resp.sendError(HttpServletResponse.SC_NOT_FOUND);
022     }
023 }
```

Figure 5.18 The `doGet()` method.

Figure 5.18. Because this servlet only supports HTTP gets, this method serves as the primary entry point into the servlet.

The first order of business is to query the `ViewPage` parameter to obtain the internal name of the page to be viewed (line 004). If this parameter has a value, the data member containing `Page` objects by name is queried (line 008). If the value of the parameter is null, however, the default page is used as stipulated in the servlet design (line 012).

NOTE

Unlike the DumpServlet from the previous example, there is no need to establish a convenience-member variable for the `HttpSession`, because the `PageView` servlet does not need to maintain a session.

Before the process of displaying the page can be initiated, you must perform one final check. Because a bogus page name might have been provided as the value of the `ViewPage` parameter, the page object might be `null`. If the reference is not `null`, the content type for the response is established as WML (line 016), and the `printPage()` method is invoked (line 017). If the page reference is `null`, an HTTP error of `SC_NOT_FOUND` is sent back to the requesting client.

The printPage( ) Method

With a great deal of preliminary work out of the way, it is now time to render a WML deck that represents a page. The implementation of the `printPage()` method is shown in Figure 5.19. This method renders the preamble, WML deck, and the first card

```
001 public void printPage(Page page, PrintWriter pw)
002 {
003     String strTitle = page.getTitle();
004     String strHeading = page.getHeading();
005     String strLinkHeading = page.getLinkHeading();

006     Enumeration enumTextLines = page.getTextLines();
007     Enumeration enumLinks = page.getLinks();

008     boolean bHasText = (enumTextLines != null);
009     boolean bHasLinks = (enumLinks != null);

010     pw.println("<?xml version=\"1.0\"?>");
011     pw.println("<!DOCTYPE wml PUBLIC \"-//WAPFORUM//DTD WML 1.1//EN\"");
012     pw.println("\"http://www.wapforum.org/DTD/wml_1.1.xml\">");

013     pw.println("<wml>");

014     pw.println("<card title=\"" + strTitle + "\">");

015     if (bHasLinks)
016     {
017         pw.println("<do type=\"accept\" label=\"Links\">");
018         pw.println("<go href=\"#links\"/>");
```

(continues)

Figure 5.19 The `printPage()` method.

```
019         pw.println("</do>");
020     }

021     pw.println("<do type=\"prev\" label=\"Back\">");
022     pw.println("<prev/>");
023     pw.println("</do>");

024     pw.println("<p align=\"center\"><b>");
025     pw.println(strHeading);
026     pw.println("</b></p>");

027     if (bHasText)
028     {
029         pw.println("<p>");

030         while (enumTextLines.hasMoreElements())
031         {
032             pw.println(enumTextLines.nextElement());
033         }

034         pw.println("</p>");
035     }

036     pw.println("</card>");

037     if (bHasLinks)
038     {
039         printLinkCard(pw, strLinkHeading, enumLinks);
040     }

041     pw.println("</wml>");

042     pw.close();
043 }
```

Figure 5.19 The `printPage()` method (Continued).

(the description card). This method also renders linkage to the second card (the links card) if the page at hand has links.

The process of rendering the WML content commences with the initialization of some variables that will be useful in processing the card. First, the title (line 003), page heading (line 004), and links heading (line 005) are obtained from the `page` object. Next, the enumerations containing the description's text lines (line 006) and links (line 007) are acquired. Finally, the value of the convenience flags `bHasText` (line 008) and `bHasLinks` (line 009) are calculated. If *true*, the `bHasText` flag indicates that the page contains text to be rendered on the first card. If *true*, the `bHasLinks` flag indicates that the page has links to other pages and/or WML decks. Once the variable set for the method has been initialized, the preamble of the WML deck is rendered (lines 010–013).

At this point, we are ready to render the first card. The `<card>` element is written by using the page title as the value of the `title` attribute (line 014). If the page has links (according to the value of the flag `bHasLinks`), the *action* soft key is associated with the internal hypertext reference `links` (lines 17–19).

NOTE

The internal hypertext reference `links` is used as the ID attribute of the second card (the card containing the links for the page).

The next step in rendering the first card consists of establishing the *prev* soft key (lines 021–023). Next, the page heading is displayed as center-aligned, bold text (lines 024–026). If the page contains text (as indicated by the `bHasText` flag), then execution of the `printPage()` method continues with the rendering of a `<p>` element consisting of the concatenated text lines (lines 027–035). Finally, the rendering of the first card concludes with the closing of the `<card>` element (line 036).

Only a few steps remain in the processing of a `PageView` servlet request. If the page has links (as indicated by the `bHasLinks` flag), the `printLinkCard()` method is invoked in order to render the second card containing the links (lines 037—040). Next, the `<wml>` element is closed (line 041). Finally, the `PrintWriter` object is closed in order to flush the content (line 042).

The printLinkCard( ) Method

As indicated previously, the `printLinkCard()` method is used to render the second card representing a page (the links card). The implementation of the method is shown in Figure 5.20. Because this method is only invoked if a page contains links, it must stand alone with respect to the syntax of the WML DTD. Therefore, it contains nothing more than a single `<card>` element.

The implementation of the method begins with the opening of the `<card>` element (line 004). As indicated earlier, the `links` internal hypertext reference is used as the value of the `id` attribute. Once the card is open, the links card heading is displayed as left-aligned, bold text (line 005). Because the balance of the card (the actual links) is implemented as a series of hypertext links, a paragraph (`<p>` element) must be opened to contain the `<a>` elements (line 006).

With the groundwork of the method complete, you can start rendering the `<a>` elements that represent the links. Because the links are accessed via an `Enumeration` object, a loop is established to process the rendering operation (lines 007–022). Each pass through the loop begins with obtaining the next `PageLink` object (line 009). Next, the label (line 10) and the link (line 11) values are obtained. If the link value starts with a pound sign (#), an internal page is indicated. In this case, an `<a>` element is rendered with a URL relative to the servlet that includes a `ViewPage` attribute with a value of the link (excluding the pound sign). This rendering is represented in lines 014–015. If the link value does not start with a pound sign, an external URL is indicated. In this case, an

```
001 protected void printLinkCard(PrintWriter pw,
002     String strHeading, Enumeration enumLinks)
003 {
004     pw.println("<card id=\"links\">");
005     pw.println("<p><b>" + strHeading + "</b></p>");
006     pw.println("<p>");

007     while (enumLinks.hasMoreElements())
008     {
009         PageLink pageLink = (PageLink) enumLinks.nextElement();

010         String strLabel = pageLink.getLabel();
011         String strLink = pageLink.getLink();

012         if (strLink.startsWith("#"))
013         {
014             pw.print("<a href=\"?ViewPage=" +
015                 strLink.substring(1) + "\">");
016         }
017         else
018         {
019             pw.print("<a href=\"" + strLink + "\">");
020         }

021         pw.println(strLabel + "</a>");
022     }

023     pw.println("</p>");
024     pw.println("</card>");
025 }
```

Figure 5.20 The `printLinkCard()` method.

`<a>` element is rendered with the absolute URL indicated (line 019). In either case, the label is rendered as the content of the `<a>` element.

Once the enumeration has been exhausted and all links have been rendered, the `<p>` element is closed (line 023). At this point, the work of the `printLinkCard()` method has concluded, so the `<card>` element is closed (line 024).

Execution

To see the `PageViewer` servlet in action, you can use the servlet engine from the accompanying CD-ROM in this book. To execute the server, you will need to perform the following procedures:

- Change directories to the `/code` directory on the CD-ROM or to a directory to which you copied the contents for experimentation purposes.

- Make sure that the directory " . " is in your CLASSPATH environment variable.
- Make sure that the `java.exe` executable from the JDK is in your execution search path.
- Enter `java com.n1guz.ServletEngine.ServletEngine` at the command line.

TIP

Because the `PageView` servlet is Pure Java, you should be able to run it from any HTTP server that supports JSDK. In order for the servlet to work correctly, however, you will need to make sure that the initialization parameter datafile is correctly established.

Once the server is online, you can start your favorite WAP browser simulator and execute the following URL to see the servlet: http://localhost/PageViewer. Following the same run-time examples shown here will yield the same results.

Conclusion

In this chapter, we learned about serving and rendering WML content for the purpose of presenting a homepage. Because this book is devoted to writing WAP servlets, we minimized the introduction of technologies beyond this realm. For example, the initialization portion of the rendering example was limited to simple text-file input/output (I/O). The techniques shown in this example could certainly be combined with other facilities of the Java language and related packages. For example, you could use the following sources for page data:

- A flat file containing an XML-based description of pages rendered by the site
- A relational database accessed via Java Database Connection (JDBC)
- Enterprise Java Beans, served by an application server

Besides these additional techniques for sourcing page data, restructuring the servlet itself could extend its functionality. For example, rather than generating page images at initialization time, pages could be rendered on the fly from live data (accessed from the aforementioned sources).

While serving homepages is a time-tested function of HTTP servers the world over, the real power of WAP Servlets can be found in their capability to serve applications that perform useful functions for the WAP browser user. In the next chapter, we will continue our evaluation of real-world applications by looking at such an example.

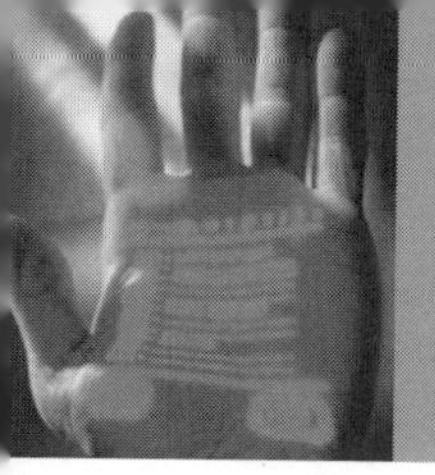

CHAPTER 6

Real-World Application Example: The Grocery Servlet

In previous chapters, we discussed *Wireless Markup Language* (WML) and its use in generating static, wireless Web content. We extended the model by exploring Sun Microsystems' *Java Servlet Development Kit* (JSDK) and by using a servlet to render static, wireless Web content on the fly. In this chapter, we will continue to build upon our understanding of *Wireless Application Protocol* (WAP) Servlets by creating the grocery servlet.

The Requirements

In Chapter 5, "*Wireless Markup Language (WML) Homepage Example*," the PageViewer servlet is not too dynamic in nature. During its initialization, the servlet reads a datafile. That datafile supplies static data used to generate static content. Such an application is good for information retrieval, but most applications need to be more interactive. Just as the wired Web moved from information retrieval to interactive sessions, the wireless Web must provide the same capabilities. The grocery servlet discussed in this chapter is our interactive example.

The goal of the grocery application is to enable users to create a shopping list. Once the list is created, the users will be able to use their WAP browsers at the grocery store while they shop. The following list describes the requirements for the application:

- There should be two main modes of operation: input and shopping.
- When in input mode, the user should be able to deal with categories of food, such as dairy and vegetables.
- When in shopping mode, users should be able to deal with the organization of their favorite stores by areas such as aisle one, delicatessen, frozen foods, and so on.

- Because people tend to buy the same items week after week, the input mode should have a function for entering a frequently purchased item and then selecting it when needed.
- When in shopping mode, the user should have an indication of which store areas contain needed items.
- When viewing items in a store area, users should only see needed items.

The Design

Based on this relatively simple list of requirements, we can begin the task of designing the application. We outline the design elements in the following list:

- When the servlet is accessed, show the user a card to choose between the two primary modes of operation. This card should contain two selections: one for stores (shopping mode) and the other for food categories (input mode).
- When users enters input mode, they should have a list of existing categories with a link for adding new categories and a link for adding new items to existing categories. For organizational purposes, the *new category* and *new item* links should appear at the bottom of this list.
- When users select the *new category* link, they should be prompted for the name of the new category. This category is then added to the category list.
- When users select the *new item* link, they should be prompted for both the name of the new item and the category to which it should be added.
- When users select an existing category, the device should display a list of items that have been assigned to that category. Users should then be able to pick items that are needed and save the category.
- When new items are added, they should be added with their select flag set. Most of the time, users will think about items as they realize that they need them.
- When users enter shopping mode, they should be shown a list of existing stores with a link for adding a new store. For organizational purposes, the *new store* link should appear at the bottom of the list.
- When the *new store* pick is selected, the user should be prompted for the name of the store.
- When a store is selected from the store list, a list of store areas should be displayed with a pick for adding new areas. For organizational purposes, the *new area* pick should appear at the bottom of the list. There should be a default store area (called "unassigned") that should also be displayed near the bottom of the list.
- When new items are added (in input mode), they should be added to the *unassigned* store area for each store.
- When users enter the *unassigned* store area, they should be shown a select list of unassigned items. As items are selected from this list, a list of the other store areas

should be shown. When a store area is selected, the unassigned item should be moved to the selected store area.

- When users select the *new area* link, they should be prompted for the name of a new store area. The new area should be added to the list of store areas. At this point, users can assign unassigned items to the new area.
- When users enter a store area, they should see a list of items in that area. Items in the area should be shown if they were picked during input mode.

Once categories, items, and areas are established, users will enter input mode to pick items that they will need during the week. At the store, users will enter shopping mode and check-off items as they place them in their shopping cart.

The Implementation

In writing this application, I chose to break down the various data items into records. For each of these records, there is both a record class and a handler class (for rendering the WML associated with the record). Table 6.1 shows a list of records, their data class, and their handler.

In addition to these classes, there are four other classes: `Grocery`, `Handler`, `Record`, and `RecordList`. The `Grocery` class is the servlet. The `Handler` class is the superclass for all data-item handlers, and the `Record` class is the superclass for all data-item records. The `RecordList` class maintains a list of `Record` objects. The next four sections describe these classes in more detail.

The Grocery Class

As stated previously, the `Grocery` class is the main class of the servlet. Figure 6.1 shows the source of the `Grocery` class. The following sections describe the major methods of the `Grocery` class. The `destroy()` and `getServletInfo()` methods are similar to implementations that we have already seen.

Table 6.1 Object Types and Their Classes

DATA ITEM	RECORD CLASS	HANDLER CLASS
User	`UserRecord`	`UserHandler`
Category	`CategoryRecord`	`CategoriesHandler`
Category Item	`ItemRecord`	`CategoryItemHandler`
Store	`StoreRecord`	`StoresHandler`
Store Area	`AreaRecord`	`AreasHandler`
Store Area Items	`ItemRecord`	`AreaItemHandler`

```
package book.chap06.Grocery;

import java.io.*;
import java.util.*;
import javax.servlet.*;
import javax.servlet.http.*;

public class Grocery extends HttpServlet
{
  protected Hashtable m_hashHandlers = null;

  public void init(ServletConfig config)
    throws ServletException
  {
    super.init(config);

    m_hashHandlers = new Hashtable();

    try
    {
      m_hashHandlers.put("User", new UserHandler());
      m_hashHandlers.put("Store", new StoresHandler());
      m_hashHandlers.put("Category", new CategoriesHandler());
      m_hashHandlers.put("CategoryItem", new CategoryItemHandler());
      m_hashHandlers.put("Area", new AreasHandler());
      m_hashHandlers.put("AreaItem", new AreaItemHandler());
    }
    catch (NoSuchMethodException e)
    {
      throw new ServletException();
    }
  }

  public void destroy()
  {
    log("terminating");
  }

  public String getServletInfo()
  {
    return "Grocery Servlet v1.0 by John L. Cook";
  }

  public void doGet(HttpServletRequest req,
    HttpServletResponse resp)
    throws ServletException, IOException
```

Figure 6.1 The `Grocery` class.

```
    {
      doPost(req, resp);
    }

    public void doPost(HttpServletRequest req, HttpServletResponse resp)
      throws ServletException, IOException
    {
      // See what the browser is willing to accept.

      String strAccept = req.getHeader("accept");

      if ((strAccept == null) || (strAccept.indexOf("wap.wml") == -1))
      {
        resp.sendError(HttpServletResponse.SC_NOT_ACCEPTABLE);
      }
      else
      {
        resp.setContentType("text/vnd.wap.wml");

        HttpSession session = req.getSession(true);

        if (session.isNew())
        {
          UserRecord userRecord = UserRecord.getUser(req);
          session.putValue("UserRecord", userRecord);

          Cookie cookie = new Cookie("UserID", userRecord.getID());
          cookie.setMaxAge(15768000); // Six months.

          resp.addCookie(cookie);
        }

        String strType = req.getParameter("Type");

        if (strType == null)
        {
          strType = "User";
        }

        Handler handler = (Handler) m_hashHandlers.get(strType);

        handler.run(req,resp,session);
      }
    }
  }
```

Figure 6.1 Continued.

The init( ) Method

In the `init()` method, the first order of business is to call the superclass's `init()` method. If you do not perform this procedure, logging will not work correctly at a minimum. Therefore, this step is essential. Next, we instantiate an instance of each of the handler derivatives. All requests will share these objects. This technique is similar to the technique used by the servlet framework itself. By avoiding instantiation and garbage collection of a new instance with each request, the servlet will run more efficiently.

The doGet( ) and doPost( ) Methods

As shown in the previous example, *Hypertext Transport Protocol* (HTTP) gets and posts are treated the same by having the `doGet()` method call the `doPost()` method. The first step is to query the HTTP header name `accept`. This header name-value pair has a value that describes the content that the browser is willing to accept. If the browser is not willing to accept WML, the method sends the error `SC_NOT_ACCEPTABLE`. This error message is friendlier than just sending WML content to a client that is not willing to accept the material.

TIP

If you are planning on writing a servlet that is capable of processing both HTML and WML content, you could query the `accept` HTTP name-value pair to determine the type of browser that is accessing the servlet.

If the browser is willing to accept WML, the content type is set to `text/vnd.wap. wml`. Because the servlet is interactive, state must be maintained in a server session object; therefore, the next order of business is to get (or create) a session via the `getSession()` method. If the session is new, the `UserRecord` object is loaded (via a call to the static `UserRecord.getUser()` method) and assigned to the session via the `putValue()` method. (As we will see, the `UserRecord.getUser()` method queries a cookie to look up an existing user record or creates a new user if the cookie is not found.) At this time, the cookie is established (or re-established) with a six-month maximum age. Therefore, the user can go as long as six months without accessing the server before the cookie expires. Re-establishing the cookie with each new session resets the timer.

Once the session is established, the query-string parameter `Type` is retrieved. I choose this parameter to describe the type of data item with which a request is working. If the `Type` parameter is not specified on a request, I assign the default value of *User*. Next, the handler object is retrieved from the hash table created in the `init()` method, and control is passed to that method.

The Record Class

The `Record` class is reasonably straightforward. Its source appears in Figure 6.2. All records in the application have two variables in common: `m_strID` and `m_strName`. The `m_strID` variable contains an internal ID used by the application to identify the `Record`, and the `m_strName` variable contains the readable name of the `Record`. Two

```
package book.chap06.Grocery;

import java.io.*;
import java.util.*;

public abstract class Record implements Serializable
{

  /**
   * All records have a unique ID within their master list.
   * An ID uniquely identifies a record within the context of
   * its user and master list.
   */

  protected String m_strID = null;

  /**
   * All records have human readable names.
   */

  protected String m_strName = null;

  /**
   * Set the ID of a record to the given value.
   */

  public void setID(String strID)
  {
    m_strID = strID;
  }

  /**
   * Return a record's ID.
   */

  public String getID()
  {
    return m_strID;
  }

  /**
   * Set the name of a record.
   */

  public void setName(String strName)
  {
```

(continues)

Figure 6.2 The `Record` class.

```
      m_strName = strName;
    }

    /**
     * Returns the name of a record.
     */

    public String getName()
    {
      return m_strName;
    }

    /**
     * Compair a record to another by compairing the string value of
     * their ID.
     */

    public int compareTo(Record record)
    {
      String strID1 = getID();
      String strID2 = record.getID();

      return strID1.compareTo(strID2);
    }
  }
```

Figure 6.2 The `Record` class (Continued).

convenience methods are defined to respectively get and set the value of the ID: `getID()` and `setID()`. Furthermore, two additional methods are defined to respectively get and set the value of the name: `getName()` and `setName()`. Finally, the `compairTo()` method is provided to compare the ID of the `Record` at hand to that of another `Record`.

The RecordList Class

From the requirements and design notes, you should see that many of the `Record` class derivatives maintain lists of other records. The `RecordList` class provides a general-purpose list mechanism for the servlet. Figure 6.3 shows the source-code listing of the class.

The `RecordList` class stores references to the records that it contains in a `java.util.Hashtable` and in a `java.util.Vector`. The `Hashtable` provides the capability for rapid lookup of records by ID. The vector provides an ordered list of records. The methods of the `RecordList` class emulate a mixture of `Hashtable` and `Vector` methods (in most cases, by actually calling those methods).

The `put()` method is used to put a record into the list. If a record has not previously been assigned an ID (either by being placed in another list or by explicit assignment), it

```
package book.chap06.Grocery;

import java.io.*;
import java.util.*;

public class RecordList implements Serializable
{
  protected int m_nNextRecordID = 0;
  protected Hashtable m_hashRecords = new Hashtable();
  protected Vector m_vOrdered = new Vector();

  public synchronized void put(Record record)
  {
    // See if the record has an ID.

    String strID = record.getID();

    if (strID == null)
    {
      strID = String.valueOf(m_nNextRecordID++);
      record.setID(strID);
    }

    m_hashRecords.put(strID, record);

    int nCount = m_vOrdered.size();
    boolean bInserted = false;

    for (int nIndex = 0; nIndex < nCount; nIndex++)
    {
      Record record2 = (Record) m_vOrdered.elementAt(nIndex);

      if (record.compareTo(record2) < 0)
      {
        m_vOrdered.insertElementAt(record2, nIndex);
        bInserted = true;
        break;
      }
    }

    if (!bInserted)
    {
      m_vOrdered.addElement(record);
    }
  }
```

(continues)

Figure 6.3 The `RecordList` class.

```
  public Enumeration elements()
  {
    return m_vOrdered.elements();
  }

  public int size()
  {
    return m_vOrdered.size();
  }

  public Record elementAt(int nIndex)
  {
    return (Record) m_vOrdered.elementAt(nIndex);
  }

  public Record get(String strID)
  {
    return (Record) m_hashRecords.get(strID);
  }

  public Record remove(String strID)
  {
    Record record = (Record) m_hashRecords.remove(strID);
    m_vOrdered.remove(record);
    return record;
  }
}
```

Figure 6.3 The `RecordList` class (Continued).

is assigned an ID during the put process. A reference to the record is placed in the hash table with its ID as a key. The reference is also inserted into the vector in order, based on the value of its ID.

The `size()` method returns the size of the `RecordList`. The `elementAt()` method returns the record at a particular index within the vector. The `get()` method returns a record reference by ID. Finally, the `remove()` method removes the record (specified by ID) from the list.

The Record Derivatives

As shown in Table 6.1, there are five record classes associated with the servlet: `AreaRecord`, `CategoryRecord`, `ItemRecord`, `StoreRecord`, and `UserRecord`. Each of these classes is extremely simple. They primarily consist of convenience methods for accessing data members.

The StoreRecord Class

The StoreRecord class contains two RecordList data members: m_areas and m_unassignedItems. The m_areas list contains AreaRecord references representing the areas of the store. The m_unassignedItems list contains ItemRecord references representing items that have not yet been assigned to a store area. This class consists of convenience methods for accessing these two lists. A prototype of the StoreRecord class appears in Figure 6.4.

The AreaRecord Class

The AreaRecord class contains one RecordList data member: m_items. The m_items list contains ItemRecord references representing the items that have been assigned to the area. This class consists of convenience methods for accessing the list. A prototype of the AreaRecord class appears in Figure 6.5.

```
public class StoreRecord extends Record implements Serializable
{
  protected RecordList m_areas = new RecordList();
  protected RecordList m_unassignedItems = new RecordList();
  public void addArea(AreaRecord area);
  public int getAreaCount();
  public Enumeration getAreas();
  public AreaRecord getArea(String strID);
  public void addUnsignedItem(ItemRecord item);
  public int getUnassignedItemCount();
  public Enumeration getUnassignedItems();
  public ItemRecord getUnassignedItem(String strID);
  public void removeUnassignedItem(String strID);
}
```

Figure 6.4 The StoreRecord class.

```
public class AreaRecord extends Record implements Serializable
{
  protected RecordList m_items = new RecordList();
  public void addItem(ItemRecord item)
  public Enumeration getItems()
  public ItemRecord getItem(String strID)
  public int getItemCount()
}
```

Figure 6.5 The AreaRecord class.

The ItemRecord Class

The ItemRecord class contains a boolean value that indicates whether the item has been picked: m_bPicked. This class has convenience methods for setting and accessing this data member. If the user selects an item for purchase at the grocery store, its m_bPicked member will be *true*; otherwise, the value will be *false*. A prototype of the ItemRecord class appears in Figure 6.6.

The CategoryRecord Class

The CategoryRecord class contains one RecordList data member: m_items. The m_items list contains the ItemRecord reference, representing the items that have been assigned to the category. This class consists of convenience methods for accessing the list. A prototype of the CategoryRecord class appears in Figure 6.7.

The UserRecord Class

The UserRecord class contains three RecordList data members: m_stores, m_categories, and m_items. The m_stores list contains StoreRecord references representing the stores that the user adds. The m_categories list contains CategoryRecord references representing the categories that the user added. The m_items list contains ItemRecord references representing the master list of items that the user added. For the most part, the methods of the class consist of convenience methods for accessing the lists, as shown in the class prototype in Figure 6.8.

```
public class ItemRecord extends Record implements Serializable
{
  protected boolean m_bPicked = true;
  public void setPicked(boolean bPicked)
  public boolean isPicked()
}
```

Figure 6.6 The ItemRecord class.

```
public class CategoryRecord extends Record implements Serializable
{
  protected RecordList m_items = new RecordList();
  public void addItem(ItemRecord item)
  public Enumeration getItems()
  public ItemRecord getItem(String strID)
  public boolean hasItems()
}
```

Figure 6.7 The CategoryRecord class.

```
public class UserRecord extends Record implements Serializable
{
  protected RecordList m_stores = new RecordList();
  protected RecordList m_categories = new RecordList();
  protected RecordList m_items = new RecordList();

  public static UserRecord getUser(HttpServletRequest req)
    throws IOException

  protected static String getUserIDFromCookie(HttpServletRequest req);

  public void addStore(StoreRecord store)
  public int getStoreCount()
  public Enumeration getStores()
  public StoreRecord getStore(String strID)
  public int getCategoryCount()
  public void addCategory(CategoryRecord category)
  public Enumeration getCategories()
  public CategoryRecord getCategory(String strID)
  public void addItem(ItemRecord item)
  public Enumeration getItems()
  public ItemRecord getItem(String strID)
}
```

Figure 6.8 The `UserRecord` class.

There are, however, two static methods of the `UserRecord` class that we must explore in more detail: `getUser()` and `getUserIDFromCookie()`. The `getUserIDFromCookie()` method's implementation appears in Figure 6.9. This method walks through the array of `Cookie` references that are retrieved from the `HttpServletRequest` (lines 008–016). When the UserID cookie is found, its value is returned as the ID of the user (lines 011–018).

The `getUser()` method returns an existing or new `UserRecord`, depending on whether the client has accessed the servlet previously. The method's implementation appears in Figure 6.10. First, the `getUserIDFromCookie()` method is called to return the ID of an existing user (line 005). If an ID is returned, the data file associated with the user is loaded (lines 011–029). If an ID was not returned, a new ID is created by using the hex-string representation of the current time and a random number (lines 033–036). Next, a new `UserRecord` instance is created, assigned the ID that was generated, and saved to a data file. In either case, the `UserRecord` reference is returned (lines 037–038).

The save( ) Method

The `save()` method (shown in Figure 6.11) is used to serialize a `UserRecord` (and therefore, all records associated with a user). First, the ID of the `UserRecord` referenced passed is obtained (line 004). This ID is used to reconstruct the name of the

```
001 protected static String getUserIDFromCookie(
     HttpServletRequest req)
002 {
003  String strRetval = null;

004  Cookie [] cookies = req.getCookies();

005  if (cookies != null)
006  {
007    int nSize = cookies.length;
008    for (int nIndex = 0; nIndex < nSize; nIndex++)
009    {
010      Cookie cookie = cookies[nIndex];

011      if (cookie.getName().equals("UserID"))
012      {
013        strRetval = cookie.getValue();
014        break;
015      }
016    }
017  }
018  return strRetval;
019 }
```

Figure 6.9 The `getUserIDFromCookie()` method.

```
001 public static UserRecord getUser(HttpServletRequest req)
002   throws IOException
003 {
004  UserRecord userRecord = null;

005  String strID = getUserIDFromCookie(req);

006  if (strID != null)
007  {
008    try
009    {
010      File file = new File(strID + ".dat");

011      if (file.exists())
012      {
013        FileInputStream fis = new FileInputStream(file);
014        BufferedInputStream bis = new BufferedInputStream(fis);
```

Figure 6.10 The `getUser()` method.

```
015         ObjectInputStream ois = new ObjectInputStream(bis);

016         userRecord = (UserRecord) ois.readObject();
017         ois.close();
018       }
019       else
020       {
021         strID = null;
022       }
023     }
024     catch (ClassNotFoundException e)
026     {
027       strID = null;
028     }
029   }

030   if (strID == null)
031   {
032     userRecord = new UserRecord();

033     long lDate = new Date().getTime();
034     double dRandom = Math.random();

035     strID = Long.toHexString(lDate) +
036       Long.toHexString(Double.doubleToLongBits(dRandom));

037     userRecord.setID(strID);

038     Handler.save(userRecord);
039   }

040     return userRecord;
041 }
```

Figure 6.10 Continued.

data file used to store the serialized data. Next, the file is opened as a `FileOutput-Stream` wrapped in both a `BufferedOutputStream` and an `ObjectOutput-Stream` (lines 005–008). Finally, the `UserRecord` object is written to the file (lines 009 and 010).

The Handler Class

As stated previously, the `Handler` class is the abstract superclass of all data item handlers in the grocery servlet. In order for the `Grocery.doGet()` function (refer to Figure 6.1) to work correctly, all handlers must implement the `run()` method; therefore,

```
001 public static void save(UserRecord userRecord)
002   throws IOException
003 {
004   String strID = userRecord.getID();

005   File file = new File(strID + ".dat");
006   FileOutputStream fos = new FileOutputStream(file);
007   BufferedOutputStream bos = new BufferedOutputStream(fos);
008   ObjectOutputStream oos = new ObjectOutputStream(bos);

009   oos.writeObject(userRecord);
010   oos.close();
011 }
```

Figure 6.11 The `save()` method.

an abstract definition of the function is provided here to force implementation in the derived classes. The `Handler` class is one of the more complex classes of the servlet, so we discuss each method individually. A prototype of the class appears in Figure 6.12.

The invoke( ) Method

Derived handlers use the `invoke()` method (shown in Figure 6.13) to invoke one of their methods from a hash table maintained by the handler. This reflection technique for invoking handler class methods is used to simplify action handling.

The printXML( ) Method

The derived handlers invoke the `printXML()` method (shown in Figure 6.14) to print the standard XML preamble that we have used in prior examples.

By calling the `printXML()` method on the PrintWriter associated with the `HttpServletResponse`, the following lines are written at the start of the generated WML document:

```
<?xml version="1.0"?>
<!DOCTYPE wml PUBLIC "-//WAPFORUM//DTD WML 1.1//EN"
"http://www.wapforum.org/DTD/wml_1.1.xml">
```

The `printXML()` method is always called immediately after obtaining the PrintWriter.

The printHeader( ) Method

The derived handlers invoke the `printHeader()` method (shown in Figure 6.15) to print WML header and standard template content. Therefore, this method writes back-to-back `<head>` and `<template>` elements.

If the `bLimitAccess` argument is *true*, an `<access>` element is written within the content of the `<head>` element. This `<access>` element limits access to referents

```
public abstract class Handler
{
  protected void invoke(Hashtable hashMethods,
    String strMethod,
    HttpServletRequest req,
    HttpServletResponse resp,
    HttpSession session)
    throws ServletException, IOException

  protected void printXML(PrintWriter pw)

  protected void printHeader(PrintWriter pw, boolean bLimitAccess,
    boolean bNoCache, boolean bRevalidate)

  protected void confirmAddition(HttpServletResponse resp,
  String strType, String strName)
  throws IOException

  protected void printConfirmation(HttpServletResponse resp,
    String strMessage)
    throws IOException

  protected void printListEntries(Enumeration enum, PrintWriter pw,
    String strType, String strIDTitle)

  protected void printListEntries(Enumeration enum, PrintWriter pw,
    String strIDTitle, String strType, String strExtraQueryPairs)

  protected void printSelectList(PrintWriter pwContent,
    String strName,
    Enumeration enum, boolean bPicked, boolean bMulti)
    throws IOException

  protected void handleAdd(PrintWriter pw, String strType)

  protected void handleAdd(PrintWriter pw, String strType,
    String strExtraQueryNVP)

  public static void save(UserRecord userRecord)
    throws IOException
}
```

Figure 6.12 The `Handler` class.

within the same domain and path as the servlet itself. Therefore, only the servlet will be capable of accessing decks that use the limit access feature (all decks except for the main menu deck). This capability prevents other WML decks and servlets from jumping into the middle of the grocery servlet.

```
001 protected void invoke(Hashtable hashMethods,
002  String strMethod,
003  HttpServletRequest req,
004  HttpServletResponse resp,
005  HttpSession session)
006  throws ServletException, IOException
007 {
008  Method method = null;

009  if (strMethod != null)
010  {
011    method = (Method) hashMethods.get(strMethod);
012  }

013  if (method != null)
014  {
015    try
016    {
017      method.invoke(this, new Object [] { req, resp, session });
018    }
019    catch (Exception e)
020    {
021      throw new ServletException();
022    }
023  }
024 }
```

Figure 6.13 The `invoke()` method.

```
protected void printXML(PrintWriter pw)
{
  pw.println("<?xml version=\"1.0\"?>");
  pw.print("<!DOCTYPE wml PUBLIC ");
  pw.println("\"-//WAPFORUM//DTD WML 1.1//EN\"");
  pw.print("\"http://www.wapforum.org/DTD");
  pw.println("/wml_1.1.xml\">");
}
```

Figure 6.14 The `printXML()` method.

NOTE

An `<access>` element could also be used on the main entry deck to limit access to the servlet to particular WML decks.

Two additional method parameters are used to control the caching of generated WML decks: `bNoCache` and `bRevalidate`. If `bNoCache` is *true*, the `<meta>` `Cache-`

```
001 protected void printHeader(PrintWriter pw, boolean bLimitAccess,
002  boolean bNoCache, boolean bRevalidate)
003 {
004  pw.println("<head>");

005  if (bLimitAccess)
006  {
007    pw.println("<access path=\"/Grocery\"/>");
008  }

009  if (bNoCache)
010  {
011    pw.print("<meta forua=\"true\" ");
012    pw.print("http-equiv=\"Cache-Control\" ");
013    pw.println("content=\"no-cache\"/>");
014    if (bRevalidate)
015    {
016      pw.print("<meta forua=\"true\" ");
017      pw.print("http-equiv=\"Cache-Control\" ");
018      pw.println("content=\"must-revalidate\"/>");
019    }
020  }
021  else
022  {
023    pw.print("<meta forua=\"true\" ");
024    pw.print("http-equiv=\"Cache-Control\" ");
025    pw.println("content=\"max-age=86400\"/>");
026  }

027  pw.println("</head>");

028  pw.println("<template>");
029  pw.println("<do type=\"prev\" label=\"Back\">");
030  pw.println("<prev/>");
031  pw.println("</do>");
032  pw.println("</template>");
033 }
```

Figure 6.15 The `printHeader()` method.

`Control` is set to *no-cache*; therefore, the WML deck will not be stored in the browser's cache. Otherwise, the `<meta> Cache-Control` is set to *max-age=86400*. Because all grocery WML decks use this method, no deck associated with the servlet will be cached for more than 24 hours. The `bNoCache` parameter is *true* for all decks that contain volatile information.

If `bNoCache` is *true*, the `bRevalidate` parameter is considered. If the `bRevalidate` parameter is *true*, the `<meta> Cache-Control` is set to *must-revalidate*. Even if a deck is not cached, it will remain on the history stack until it is backed over. Therefore, the only way to ensure that a non-cached deck is not backed into (exposing stale data) is by forcing revalidation.

These elements of cache control are important to the grocery servlet, and we will discuss them next as the `printHeader()` method is invoked.

TIP

While HTML browser users might navigate backwards as they move around a site, most HTML Web masters construct their sites so that users can navigate via links and never need to press the back button. This functionality works well on a desktop system, because these systems have a great deal of cache space. WAP browsers tend to not have much space, so it is considered bad form to overload the history stack. Therefore, you will notice that I have gone to great lengths to navigate back up the history stack as transactions are completed. Using this technique, users can navigate backwards out of the grocery application when they are done. Had I constructed the application so that the user is always moving forward (a much easier task), even a short session could cause items to fall off the history stack.

The printConfirmation( ) Method

The derived handlers invoke the `printConfirmation()` method (shown in Figure 6.16) to render a deck that confirms some user action. The deck displayed by this method establishes a 1.5-second timer, displays the specified message, and executes a `<prev>` when the timer expires. For example, this method is used by `confirmAddition()`, which is called whenever a new `Record` class derivative is added (for example, adding a store).

NOTE

The `printHeader()` method is called with `bNoCache` set to *true*. Confirmation message data (`strMessage`) is volatile.

The confirmAddition( ) Method

The derived handlers invoke the `confirmAddition()` method (shown in Figure 6.17b) to render a deck that confirms an addition of a record derivative (for example, adding an area to a store). The `strType` parameter specifies a type of record (User, Area, Category, Store, or Item). The `strName` parameter is the name of the record (for example, Vegetable, Apple, and so on). Figure 6.17a shows an example screen shot where `strType` is Category and `strName` is Fruits & Vegetables.

TIP

When you know or suspect that rendered text might contain a character that has special meaning to XML (such as <, >, ", and &), you should place the text within a CDATA. All text that comes between `<![CDATA[` and `]]` is interpreted as character data, rather than XML. So, for example, if you want to display the text `<looks`

```
001 protected void printConfirmation(HttpServletResponse resp,
002  String strMessage)
003  throws IOException
004 {
005  PrintWriter pw = resp.getWriter();

006  printXML(pw);

007  pw.println("<wml>");
008  printHeader(pw, true, true, false);

009  pw.println("<card>");

010  pw.println("<onevent type=\"onenterforward\">");
011  pw.println("<refresh>");
012  pw.println("<setvar name=\"timer\" value=\"15\"/>");
013  pw.println("</refresh>");
014  pw.println("</onevent>");

015  pw.println("<onevent type=\"onenterbackward\">");
016  pw.println("<prev/>");
017  pw.println("</onevent>");

018  pw.println("<onevent type=\"ontimer\">");
019  pw.println("<prev/>");
020  pw.println("</onevent>");

021  pw.println("<timer name=\"timer\" value=\"15\"/>");
022  pw.print("<p mode=\"wrap\">" + strMessage + "</p>");
023  pw.println("</card>");
024  pw.println("</wml>");
025  pw.close();
026 }
```

Figure 6.16 The `printConfirmation()` method.

`like="xml">` within a paragraph, you would write it as follows: `<![CDATA[<looks like="xml">]].`

The printListEntries( ) Method

The `printListEntries()` method is used to display a list of entries in a list maintained by a derived handler. The source code for the method appears in Figure 6.18b. List items are realized as hypertext links. As stated previously, I find this method to be the best way to represent lists across a wide range of WAP browsers. The `enum` parameter provides an enumeration of `Record` class derivatives. The `strType`, `strIDTitle`, and optional `strExtraQueryPairs` parameters are used to construct the target

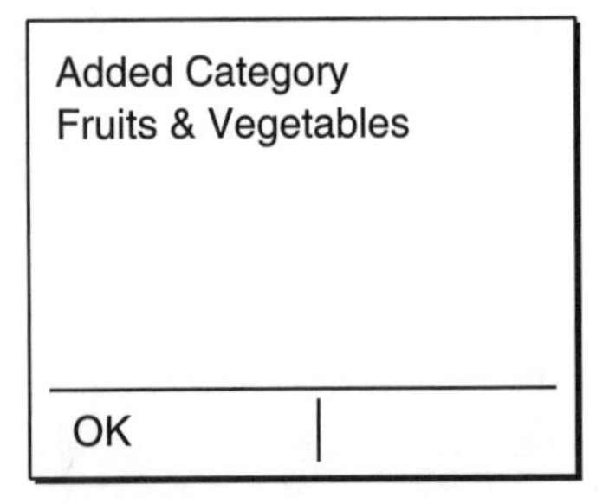

Figure 6.17a A sample `confirmAddition()` deck.

```
protected void confirmAddition(HttpServletResponse resp,
  String strType, String strName)
  throws IOException
{
  printConfirmation(resp, "<b>Added <![CDATA[" + strType + "]]></b>" +
    "<br/><![CDATA[" + strName + "]]>");
}
```

Figure 6.17b The `confirmAddition()` method.

URL. For example, the `printListEntries()` is used to display a list of areas associated with a store, as shown in Figure 6.18a (the title line comes from the `AreasHandler.handleView()` method).

The printSelectList( ) Method

The `printSelectList()` method (shown in Figure 6.19) is similar in appearance to the `printListEntries()` method; however, this method is used to collect input rather than to navigate to another WML deck. The method is flexible in the way that it does its work. The `Handler` derivative that invokes this method specifies an enumeration of `Record` references (the `enum` argument). Two additional arguments are used to control how the `<select>` element is constructed: `bPicked` and `bMulti`.

If the value of `bPicked` is *true*, each `ItemRecord` reference in the enumeration is checked. If it is picked (`isPicked()` returns *true*), an `<option>` element is included in the `<select>` element for that `ItemRecord`. Otherwise, if `bPicked` is *false*, all record references are represented by an `<option>` element. For example, when the user views a category, all items of the category are displayed (`bPicked` is *false*). Conversely, when the user views an area, only those items that are currently picked are displayed (`bPicked` is *true*).

If the value of the `bMulti` argument is *true*, the `<select>` element's `multiple` attribute is set to *true*—enabling multi-select mode. Otherwise, if the `bMulti` argument is *false*, the `<select>` element is left in single-select mode. For example, when users

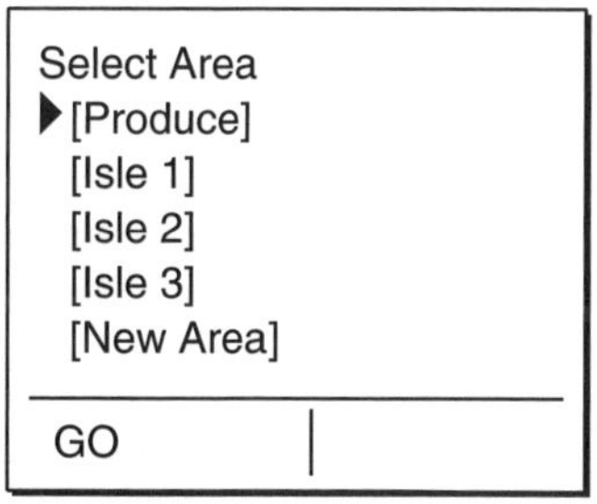

Select Area
▶ [Produce]
[Isle 1]
[Isle 2]
[Isle 3]
[New Area]

GO

Figure 6.18a A `printListEntries()` method screen shot.

```
001 protected void printListEntries(Enumeration enum, PrintWriter pw,
002   String strType, String strIDTitle)
003 {
004  printListEntries(enum, pw, strIDTitle, strType, "");
005 }

006 protected void printListEntries(Enumeration enum, PrintWriter pw,
007  String strIDTitle, String strType, String strExtraQueryPairs)
008 {
009  while (enum.hasMoreElements())
010  {
011    Record record = (Record) enum.nextElement();

012    String strName = record.getName();
013    String strID = record.getID();

014    pw.print("<p mode=\"nowrap\">");
015    pw.print("<a title=\"GO\" href=\"?Type=" + strType + "&amp;");
016    pw.print("Action=View&amp;" + strIDTitle + "=" + strID);
017    pw.print(strExtraQueryPairs);
018    pw.print("\">");
019    pw.print("<![CDATA[" + strName + "]]>");
020    pw.print("</a>");
021    pw.println("</p>");
022  }
023 }
```

Figure 6.18b The `printListEntries()` method.

view a category, they are able to select multiple items to add to their shopping list (`bMulti` is *true*). Conversely, when users assign unassigned items to a store area, they are only able to select a single item and a single area (`bMulti` is *false*).

Regardless of the settings of `bPicked` and `bMulti`, the `strName` argument is used to set up the name part of the name-value pair associated with the select (the `name` attribute of the `<select>` element). Furthermore, the ID of the records included as

```
001 protected void printSelectList(PrintWriter pwContent,
     String strName,
002  Enumeration enum, boolean bPicked, boolean bMulti)
003  throws IOException
004 {
005   boolean bEmpty = true;

006  CharArrayWriter caw = new CharArrayWriter(1024);
007  PrintWriter pw = new PrintWriter(caw);

008  pw.print("<select name=\""+ strName +"\"");

009  if (bMulti)
010  {
011    pw.print(" multiple=\"true\"");
012  }

013  pw.println(">");

014  while (enum.hasMoreElements())
015  {
016    Record record = (Record) enum.nextElement();

017    boolean bShow = true;

018    if (bPicked && (record instanceof ItemRecord))
019    {
020      ItemRecord item = (ItemRecord) record;
021      bShow = item.isPicked();
022    }

023    if (bShow)
024    {
025      bEmpty = false;
026      pw.print("<option value=\"" + record.getID() + "\">");
027      pw.print("<![CDATA[" + record.getName() + "]]>");
028      pw.println("</option>");
029    }
030  }

031  pw.println("</select>");
032  pw.close();

033  if (bEmpty)
034  {
035    pwContent.println("<i>Empty List</i>");
```

Figure 6.19 The `printSelectList()` method.

```
036   }
037   else
038   {
039     pwContent.write(caw.toCharArray());
040   }
041 }
```

Figure 6.19 Continued.

<option> elements within the <select> element is used as the value portion of the of the name-value pair (the value attribute of the <option> element).

Having a <select> element without content is illegal in this case (no <option> elements). In this method, two conditions can lead to this outcome. First, the enumeration might be empty. Second, the bPicked argument is *true*, and no ItemRecord existing within the enumeration is picked. In either case, the <select> element is replaced by a <p> element with text indicating that the list is empty. Rather than making two passes through the enumeration (first to determine whether there are any <option> elements to be generated, and second to generate them), I print the <select> element and its potential <option> elements to a CharArrayWriter and later print the contents of the writer to the PrintWriter provided an <option> element is written.

Unlike printListEntries(), this method does not create any hypertext links. Rather, it is establishing a WML variable that can be used by the calling handler in the construction of a hypertext link.

The handleAdd( ) Method

The handleAdd() method (shown in Figure 6.20b) is a general-purpose method for displaying a form to collect the name of a new record. The need to create a record derivative with a specified name is a common occurrence within the servlet. StoreRecord, CategoryRecord, AreaRecord, and ItemRecord instances are all created by using this technique. In each case, a deck is displayed with a link for creating a new record. This link will invoke the appropriate handler that will, in turn, call this method. Once the form generated by this method is submitted, the appropriate handler will be invoked, and it will confirm the addition by invoking the confirmAddition() method. Figure 6.20a shows a progression of screen shots associated with adding a store.

The first panel in Figure 6.20a shows a WML deck displaying a list of stores. When the New Store is selected, the handleAdd() method is invoked by the StoresHandler (in this case) to display a form for entering a new store's name, as shown in the second panel. Once the *accept* soft key is pressed, the WML deck resulting from the handler's

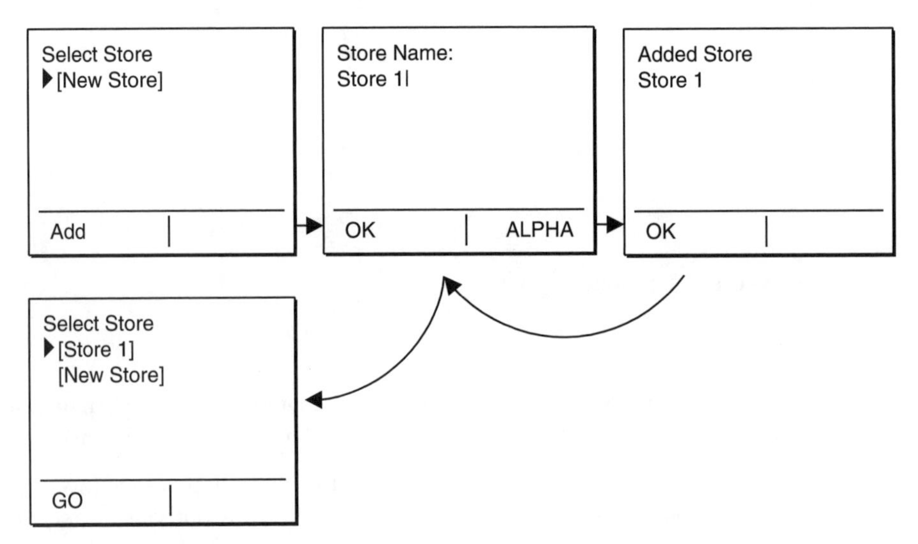

Figure 6.20a Screen shot progression for the `handleAdd()` method.

```
001 protected void handleAdd(PrintWriter pw, String strType)
002 {
003  handleAdd(pw,strType,null);
004 }

005 protected void handleAdd(PrintWriter pw, String strType,
006   String strExtraQueryNVP)
007 {
008  printXML(pw);

009  pw.println("<wml>");
010  printHeader(pw, true, false, false);

011  pw.println("<card>");

012  pw.println("<onevent type=\"onenterforward\">");
013  pw.println("<refresh>");
014  pw.println("<setvar name=\"Name\" value=\"\"/>");
015  pw.println("</refresh>");
016  pw.println("</onevent>");

017  pw.println("<onevent type=\"onenterbackward\">");
018  pw.println("<prev/>");
019  pw.println("</onevent>");
```

Figure 6.20b The `handleAdd()` method.

```
020  pw.println("<do type=\"accept\">");

021  pw.print("<go href=\"?Type=" + strType + "&amp;");
022  pw.print("Action=Set");

023  if (strExtraQueryNVP != null)
024  {
025    pw.print(strExtraQueryNVP);
026  }

027  pw.print("&amp;Name=$(Name)");
028  pw.println("\"/>");

029  pw.println("</do>");
030  pw.println("<p>");
031  pw.println("<b>" + strType + " Name:</b>");
032  pw.println("<input type=\"text\" name=\"Name\"/>");
033  pw.println("</p>");
034  pw.println("</card>");
035  pw.println("</wml>");
036 }
```

Figure 6.20b Continued.

calling `confirmAddition()` is shown in panel three. Recalling our previous discussion of `printConfirmation()`, once the timer expires, a `<prev/>` will be executed —taking us back to the previous card in the history stack of the browser. Because we do not want to see this card again, the `handleAdd()` method rendered a `<onevent>` element of type *onenterbackward* with an action of `<prev/>` (lines 17–19) in Figure 6.20b. Thus, we will skip back on the history stack to the first deck. Because this deck was rendered with no-cache and revalidate cache controls, it will be fetched again to reveal the new addition, as shown in panel four.

NOTE

While `handleAdd()` could mark the deck that it renders for revalidation, it is better if you do not perform this action. Because we know that it has the *onenterbackwards* type, resulting in a `<prev/>`, we can simply use the stale version of the non-cached deck on the history stack as a springboard back to the previous card—thereby avoiding a needless network transaction. The bottom line is that you should neither ignore nor always use the revalidate cache-control.

You should also take note that the `handleAdd()` method always clears the Name variable upon forward entry into its `<card>` element (lines 12–16). Although the deck itself is not cached, the value of the `$(Name)` variable is persistent across decks.

The Handler Derivative Classes

Before diving into the individual class derived from the Handler class, we should begin by looking at what is common between the classes in terms of functionality. To assist with this effort, Figure 6.21 shows the classes and their methods.

As you can see, all classes have a run() method. As discussed earlier, the run() method is used to invoke an action based on a Hashtable of actions maintained by each Handler derivative class via the Handler.invoke() method. For example, Figure 6.22 shows the constructor and run() method of the CategoryItemHandler class.

As shown on line 1, the Handler classes have a static data member for holding an action dispatch table. If the member has not been initialized on a previous instantiation of the CategoryItemHandler, it is initialized in lines 004–022. First, a parameter signature required by all action-handler routines is created in lines 007–012. Next, for each action handler, the reflection method (with a matching signature) is placed in the hash table with a well-known action name as its key (lines 014–021).

When the run() method is invoked, it looks up the well-known name passed as the HTTP parameter called Action (line 029). Next, the previously established reflection method with the matching name is invoked via the Handler.invoke() method (line 030).

The constructor and run() method of each of the Handler derivatives are essentially the same; therefore, we will not discuss them individually. Instead, refer to the sources on the accompanying CD-ROM if you would like to see the individual implementations. Table 6.2 provides a list of all Handler classes, their handler methods, and the associated well-known name as a reference aide.

Another common attribute of most of the Handler classes is that they have a handleAdd() and handleSet() method. In most cases, the handleAdd() method sim-

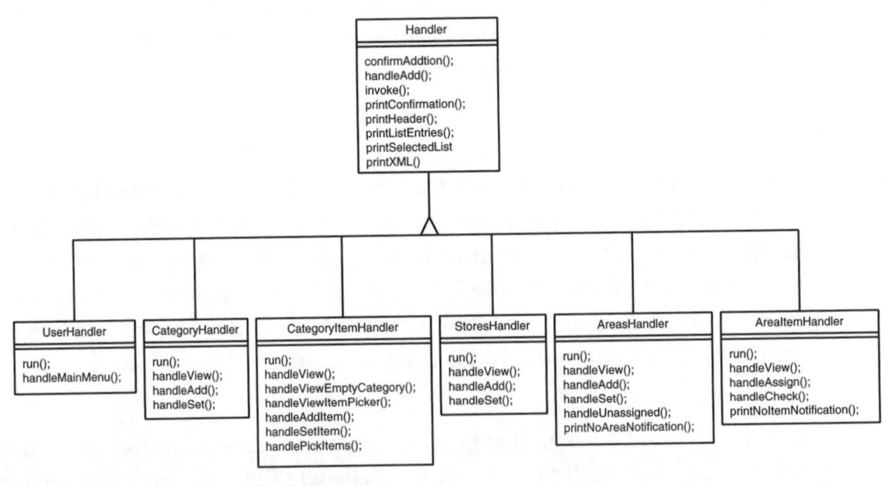

Figure 6.21 Handler class hierarchy.

```
001 protected static Hashtable sm_hashActions = null;

002 public CategoryItemHandler() throws NoSuchMethodException
003 {
004  if (sm_hashActions == null)
005  {
006    Class theClass = CategoryItemHandler.class;

007    Class signature[] = new Class []
008    {
009      HttpServletRequest.class,
010      HttpServletResponse.class,
011      HttpSession.class
012    };

013    sm_hashActions = new Hashtable();

014    sm_hashActions.put("View",
015      theClass.getMethod("handleView", signature));

016    sm_hashActions.put("Add",
017      theClass.getMethod("handleAdd", signature));

018    sm_hashActions.put("Set",
019      theClass.getMethod("handleSet", signature));

020    sm_hashActions.put("PickItems",
021     theClass.getMethod("handlePickItems", signature));
022  }
023 }

024 public void run(HttpServletRequest req,
025  HttpServletResponse resp,
026  HttpSession session)
027  throws ServletException, IOException
028 {
029  String strAction = req.getParameter("Action");

030  invoke(sm_hashActions, strAction, req, resp, session);
031 }
```

Figure 6.22 An example `run()` method.

ply invokes the `Handler.handleAdd()` with appropriate parameters. The `handleSet()` method is the target destination of the *accept* `<do>` of the WML card generated by the matching `handleAdd()`. For an example of a `handleAdd()/handleSet()` pair, refer to the methods from the `StoresHandler` class (shown in Figure 6.23).

Table 6.2 `Handler` Derivative Reflection Methods

CLASS	WELL-KNOWN NAME	METHOD
UserHandler	MainMenu	handleMainMenu()
CategoriesHandler	View	handleView()
	Add	handleAdd()
	Set	handleSet()
CategoryItemHandler	View	handleView()
	Add	handleAdd ()
	Set	handleSet ()
	PickItem	handlePickItem()
StoresHandler	View	handleView
	Add	handleAdd
	Set	handleSet
AreasHandler	View	handleView
	Add	handleAdd
	Set	handleSet
	Unassigned	handleUnassigned
AreaItemHandler	View	handleView
	Assign	handleAssign
	Check	handleCheck

As stated earlier, the `handleAdd()` method invokes the `Handler.handleAdd()` method previously discussed (line 007). As you will recall, the `Store` parameter is used to both display a prompt (Store Name) and to represent the value for the `Type` parameter passed in the request resulting from the *accept* <do>.

The `handleSet()` method (lines 10–30 in Figure 6.23) shows how the results from the form rendered by `handleAdd()` are processed. The `Name` parameter is retrieved in line 015. Next, the `UserRecord` associated with the current user is retrieved from the session (line 016–017). Then, a new `StoreRecord` is created (line 020) and assigned a name (line 021). Finally, the store is added to the user record; the user record is saved; and a confirmation message is sent to the browser (lines 022–024). If the `UserRecord` is not found (implying that something is wrong or that the user is attempting to tamper with the servlet), the error code `SC_BAD_REQUEST` is returned.

The `handleAdd()`/`handleSet()` pairs from the other `Handler` derivatives are functionally the same; thus, we will not discuss them individually. Refer to the sources on the accompanying CD-ROM for individual implementations.

```
001 public void handleAdd(HttpServletRequest req,
002  HttpServletResponse resp,
003  HttpSession session)
004  throws ServletException, IOException
005 {
006  PrintWriter pw = resp.getWriter();
007  super.handleAdd(pw, "Store");
008  pw.close();
009 }

010 public void handleSet(HttpServletRequest req,
011  HttpServletResponse resp,
012  HttpSession session)
013  throws ServletException, IOException
014 {
015  String strName = req.getParameter("Name");

016  UserRecord userRecord =
017    (UserRecord) session.getValue("UserRecord");

018  if (userRecord != null)
019  {
020    StoreRecord storeRecord = new StoreRecord();
021    storeRecord.setName(strName);

022    userRecord.addStore(storeRecord);

023    save(userRecord);
024    confirmAddition(resp, "Store", strName);
025  }
026  else
027  {
028    resp.sendError(HttpServletResponse.SC_BAD_REQUEST);
029  }
030 }
```

Figure 6.23 Example `handleAdd()` and `handleSet()` methods.

The UserHandler Class

The primary function of the `UserHandler` class (see Figure 6.24) is to process requests associated with users themselves. The `UserHandler` class is invoked by `Grocery.doPost()` in response to having received a request containing the parameter `Type=User`. Furthermore, if no `Type` parameter is specified within the query string, the `UserHandler` is the default handler. Table 6.3 shows a table of query string parameters that are used to invoke the class instance and a description of the action taken.

```
package book.chap06.Grocery;

public class UserHandler extends Handler
{
  public UserHandler()
    throws NoSuchMethodException;
  public void run(HttpServletRequest req,
    HttpServletResponse resp,
    HttpSession session)
    throws ServletException, IOException;
  public void handleMainMenu(HttpServletRequest req,
    HttpServletResponse resp,
    HttpSession session)
    throws ServletException, IOException;
}
```

Figure 6.24 `UserHandler` prototype.

Table 6.3 `UserHandler` Query Strings

QUERY STRING	DESCRIPTION
n/a	Displays the servlet's main menu
`?Type=MainMenu`	Displays the servlet's main menu

The handleMainMenu( ) Method

The only atypical method of the `UserHandler` class is the `handleMainMenu()` method (see Figure 6.25b). If the grocery servlet is entered via the URL http://localhost/Grocery (in other words, without a query string), the response will be rendered by this method. The WML deck rendered by this method is also unique in that it is the only non-volatile deck. Thus, caching is allowed (line 009). A hypertext link is established for navigating to the user's list of stores (lines 012–015). Another link is established for navigating to the user's list of categories (lines 016–019). Figure 6.25a shows a screen shot of the resulting deck.

The CategoriesHandler Class

The primary function of the `CategoriesHandler` class is to process requests associated with the collection of categories belonging to the requesting user. An instance of the `CategoriesHandler` is invoked by the `Grocery.doPost()` method in response to having received a request containing the parameter `Type=Category`. Table 6.4 shows a table of query string parameters that are used to invoke the class instance and a description of the action taken.

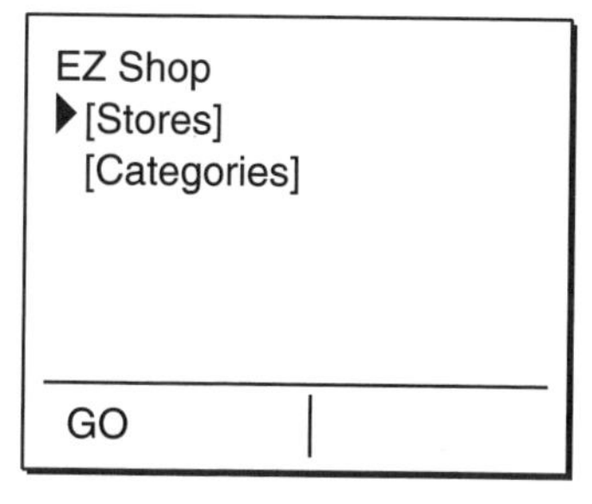

Figure 6.25a Example main menu.

```
001 public void handleMainMenu(HttpServletRequest req,
002  HttpServletResponse resp,
003  HttpSession session)
004  throws ServletException, IOException
005 {
006  PrintWriter pw = resp.getWriter();

007  printXML(pw);

008  pw.println("<wml>");
009  printHeader(pw, false, false, false);

010  pw.println("<card>");

011  pw.println("<p><b>EZ Shop</b></p>");

012  pw.print("<p><a title=\"GO\" href=\"");
013  pw.print("?Type=Store");
014  pw.print("&amp;Action=View");
015  pw.println("\">Stores</a></p>");

016  pw.print("<p><a title=\"GO\" href=\"");
017  pw.print("?Type=Category");
018  pw.print("&amp;Action=View");
019  pw.println("\">Categories</a></p>");

020  pw.println("</card>");

021  pw.println("</wml>");
022  pw.close();
023 }
```

Figure 6.25b The `UserHandler.handleMainMenu()` method.

Table 6.4 `CategoriesHandler` Query Strings

QUERY STRING	DESCRIPTION
?Type=Category&Action=View	Displays a list of categories associated with a user
?Type=Category&Action=Add	Displays a form for adding a new category
?Type=Category&Action=Set&Name=<name>	Establishes a new category

```
package book.chap06.Grocery;

public class CategoriesHandler extends Handler
{
  public CategoriesHandler()
    throws NoSuchMethodException;
  public void run(HttpServletRequest req,
    HttpServletResponse resp,
    HttpSession session)
    throws ServletException, IOException;
  public void handleView(HttpServletRequest req,
    HttpServletResponse resp,
    HttpSession session)
    throws ServletException, IOException;
  public void handleAdd(HttpServletRequest req,
    HttpServletResponse resp,
    HttpSession session)
    throws ServletException, IOException;
  public void handleSet(HttpServletRequest req,
    HttpServletResponse resp,
    HttpSession session)
    throws ServletException, IOException;
}
```

Figure 6.26 `CategoriesHandler` prototype.

The handleView( ) Method

The `handleView()` method of the `CategoriesHandler` class (refer to Figure 6.27b) is responsible for rendering a deck that shows the categories associated with a user. Due to the volatility of the deck in both the forward and backward directions, this method enables no-cache and revalidation (line 013). The `printListEntries()` method of the `Handler` class is used to display the enumeration of categories as a list of hypertext links (lines 016–017). Next, a link is added after the list for adding a new category (lines 018–021). Finally, if there are categories, a link is added to the end of the card for adding an item (lines 022–028).

Figure 6.27a shows a progression of screen shots. The first panel shows the main menu. Once the categories link is selected, this method displays the WML document shown in the second panel. (Note that the *Add Item* link is not rendered, because no categories have been added.) The third panel shows the Add Category panel rendered by the `handleAdd()` method of the `CategoriesHandler` class. The fourth panel shows the add acknowledgement, and finally, the fifth panel shows the results of calling this method again. Although the acknowledgement deck backs up the history stack to the view categories deck, because the view categories deck has a revalidate cache-control, it is reloaded. Thus, the newly added category is shown. Furthermore, because there is a category, the add item link is shown.

The CategoryItemHandler Class

The primary function of the `CategoryItemHandler` class is to process requests associated with a specific category or its items. An instance of the `CategoryItemHandler` is invoked by the `Grocery.doPost()` method in response to having received a request containing the parameter `Type=CategoryItem`. Table 6.5 shows a table of query string parameters that are used to invoke the class instance and a description of the action taken.

The handleView( ) Method

The `handleView()` method of the `CategoryItemHandler` class (refer to Figure 6.29) is used to render the items in a category so that a user can pick items that he or she would like to purchase at the store. The first task is to obtain the category ID (line 006) and establish values for the `UserRecord` and `CategoryRecord` (lines 007–018). If the category contains items, the `handleViewItemPicker()` is invoked to render the list of items (line 31). Otherwise, the `handleViewEmptyCategory()` method is called to show a message instructing the user to enter an item (line 35). Due to the volatility of the deck in both the forward and backward directions, this method enables no-cache and revalidation (line 024).

The handleViewEmptyCategory( ) Method

The `handleViewEmptyCategory()` method (shown in Figure 6.30b) displays a simple message instructing the user to add an item to a category before accessing the item. Figure 6.30a shows a screen shot of the deck generated by calling this method.

The handleAdd( ) Method

The `handleAdd()` method (shown in Figure 6.31b) is slightly different from the typical add method shown earlier. Therefore, it warrants a brief discussion. Besides the typical setup (lines 006–030) and the `<input>` element for prompting the user for the item

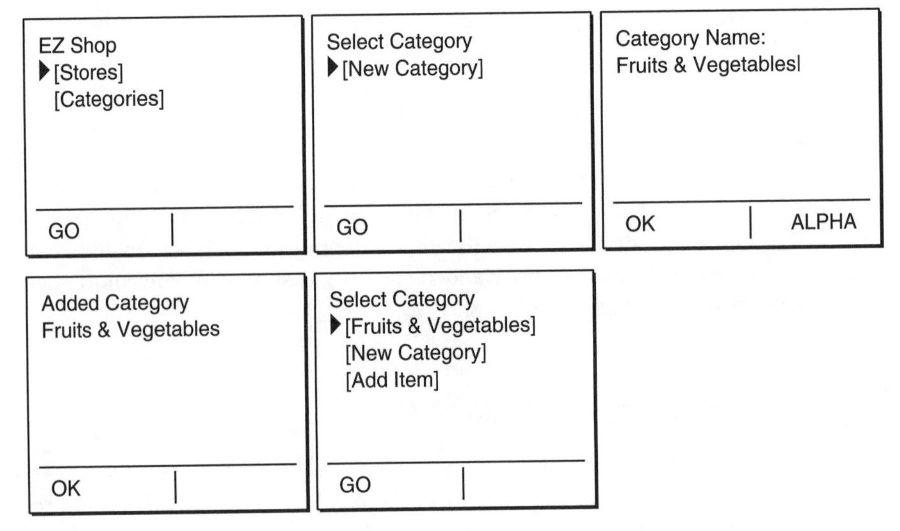

Figure 6.27a Example categories view.

```
001 public void handleView(HttpServletRequest req,
002  HttpServletResponse resp,
003  HttpSession session)
004  throws ServletException, IOException
005 {
006  UserRecord userRecord =
007    (UserRecord) session.getValue("UserRecord");

008  if (userRecord != null)
009  {
010    PrintWriter pw = resp.getWriter();

011    printXML(pw);

012    pw.println("<wml>");
013    printHeader(pw, true, true, true);

014    pw.println("<card>");

015    pw.println("<p><b>Select Category</b></p>");

016    Enumeration enum = userRecord.getCategories();
017    printListEntries(enum,pw,"CategoryItem","Category");

018    pw.print("<p><i><a title=\"Add\" href=\"");
```

Figure 6.27b The `CategoriesHandler.handleView()` method.

```
019     pw.print("?Type=Category");
020     pw.print("&amp;Action=Add");
021     pw.println("\">New Category</a></i></p>");

022     if (userRecord.getCategoryCount() > 0)
023     {
024       pw.print("<p><i><a href=\"");
025       pw.print("?Type=CategoryItem");
026       pw.print("&amp;Action=AddItem");
027       pw.println("\">Add Item</a></i></p>");
028     }
029     pw.println("</card>");

030     pw.println("</wml>");
031     pw.close();
032   }
033   else
034   {
035     resp.sendError(HttpServletResponse.SC_BAD_REQUEST);
036   }
037 }
```

Figure 6.27b Continued.

Table 6.5 `CategoryItemHandler` Query Strings

QUERY STRING	DESCRIPTION
`?Type=CategoryItem&Action =View&Category=<id>`	Displays the items within a category
`?Type=CategoryItem&Action =Add&Category=<id>`	Displays a form for adding a new item
`?Type=CategoryItem&Action =Set&Category=<id>&Name=<name>`	Establishes a new item
`?Type=CategoryItem&Action =Pick&Category=<id>&Picked=<ids>`	Marks items as wanted

name (lines 031–034), there is also a call to `printSelectList()` in the `Handler` class (line 037). This method displays a list of categories. By selecting a category, the user indicates the category that should receive the newly created item.

Figure 6.31a shows a screen shot progression demonstrating the addition of an item to a category. Panel one shows the prompt for the item name. The second panel shows the `<select>` element used to select a category. The third panel shows the add confirmation.

```
package book.chap06.Grocery;

public class CategoryItemHandler extends Handler
{
  public CategoryItemHandler()
    throws NoSuchMethodException;
  public void run(HttpServletRequest req,
    HttpServletResponse resp,
    HttpSession session)
    throws ServletException, IOException;
  public void handleView(HttpServletRequest req,
    HttpServletResponse resp,
    HttpSession session)
    throws ServletException, IOException;
  protected void handleViewEmptyCategory(PrintWriter pw,
    CategoryRecord categoryRecord);
  public void handleAdd(HttpServletRequest req,
    HttpServletResponse resp,
    HttpSession session)
    throws ServletException, IOException;
  public void handleSet (HttpServletRequest req,
    HttpServletResponse resp,
    HttpSession session)
    throws ServletException, IOException;
  protected void handleViewItemPicker(PrintWriter pw,
    String strCategory,
    CategoryRecord categoryRecord);
  public void handlePickItems(HttpServletRequest req,
    HttpServletResponse resp, HttpSession session)
    throws ServletException, IOException;
}
```

Figure 6.28 `CategoryItemHandler` prototype.

The handleViewItemPicker( ) Method

The `handleViewItemPicker()` method (shown in Figure 6.32b) is called by the `handleView()` method when there are items in the category that can be picked. This method uses a `<select>` element with the multi attribute set to *true* in order to display the items in the category (lines 41–42). The currently picked items need to be represented by the `$(Picked)` variable. Furthermore, the `<option>` elements need to be rendered. To avoid making multiple passes through the items of the category, I chose to render the `<option>` elements in a `CharArrayWriter` (lines 022–024) at the same time that I constructed the initialization string for the `$(Picked)` variable (lines 012–019). Finally, the elements of the `<card>` element are written to the passed `PrintWriter` (lines 027–042), including the contents of the `CharArrayWriter` (line 043).

Figure 6.32a shows a screen shot demonstrating the content rendered by the `handleViewItemPicker()` method. This example shows a category containing three items:

```
001 public void handleView(HttpServletRequest req,
002  HttpServletResponse resp,
003  HttpSession session)
004  throws ServletException, IOException
005 {
006  String strCategory = req.getParameter("Category");

007  CategoryRecord categoryRecord = null;
008  UserRecord userRecord = null;

009  if (strCategory != null)
010  {
011    userRecord =
012      (UserRecord) session.getValue("UserRecord");
013  }

014  if (userRecord != null)
015  {
016    categoryRecord =
017      userRecord.getCategory(strCategory);
018  }

019  if (categoryRecord != null)
020  {
021    PrintWriter pw = resp.getWriter();

022    printXML(pw);

023    pw.println("<wml>");
024    printHeader(pw, true, true, true);

025    pw.println("<card>");

026    pw.println("<onevent type=\"onenterbackward\">");
027    pw.println("<prev/>");
028    pw.println("</onevent>");

029    if (categoryRecord.hasItems())
030    {
031      handleViewItemPicker(pw, strCategory, categoryRecord);
032    }
033    else
034    {
035      handleViewEmptyCategory(pw, categoryRecord);
036    }
```

(continues)

Figure 6.29 The `CategoryItemHandler.handleView()` prototype.

```
037     pw.println("</card>");

038     pw.println("</wml>");
039     pw.close();
040   }
041   else
042   {
043     resp.sendError(HttpServletResponse.SC_BAD_REQUEST);
044   }
045 }
```

Figure 6.29 The `CategoryItemHandler.handleView()` prototype (Continued).

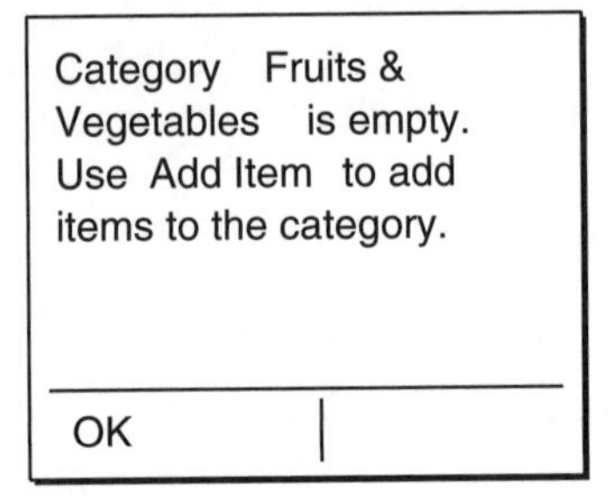

Figure 6.30a The empty category message.

```
001 protected void handleViewEmptyCategory(PrintWriter pw,
002   CategoryRecord categoryRecord)
003 {
004   String strName = categoryRecord.getName();

005   pw.println("<p><b>Category <i><![CDATA[" + strName + "]]></i>");
006   pw.println("is empty.</b></p>");
007   pw.println("<p>Use <i>Add Item</i> to add items to");
008   pw.println("the category.</p>");
009 }
```

Figure 6.30b The `handleViewEmptyCategory()` method.

apples, grapes, and carrots. In this case, both graphs and carrots have been picked; however, apples have not been picked.

The handlePickItems( ) Method

The `handlePickItems()` method (shown in Figure 6.33) is invoked in response to clicking the *accept* soft key (*save*) on the deck rendered by the `handleViewItem-Picker()` method. First, the `UserRecord` is obtained from the session (line 006).

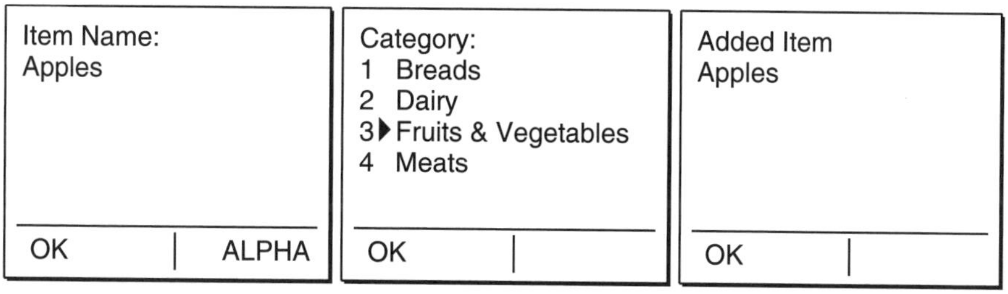

Item Name:
Apples

OK | ALPHA

Category:
1 Breads
2 Dairy
3▸Fruits & Vegetables
4 Meats

OK |

Added Item
Apples

OK |

Figure 6.31a A `handleAdd()` example.

```
001 public void handleAdd(HttpServletRequest req,
002  HttpServletResponse resp,
003  HttpSession session)
004  throws ServletException, IOException
005 {
006  String strItemName = req.getParameter("Name");
007  String strCategory = req.getParameter("Category");

008  UserRecord userRecord =
009    (UserRecord) session.getValue("UserRecord");

010  PrintWriter pw = resp.getWriter();

011  printXML(pw);

012  pw.println("<wml>");
013  printHeader(pw, true, true, false);

014  pw.println("<card>");

015  pw.println("<onevent type=\"onenterforward\">");
016  pw.println("<refresh>");
017  pw.println("<setvar name=\"Name\" value=\"\"/>");
018  pw.println("</refresh>");
019  pw.println("</onevent>");

020  pw.println("<onevent type=\"onenterbackward\">");
021  pw.println("<prev/>");
022  pw.println("</onevent>");

023  pw.println("<do type=\"accept\">");
024  pw.print("<go href=\"");
025  pw.print("?Type=CategoryItem");
```

(continues)

Figure 6.31b The `CategoryItemHandler.handleAdd()` method.

```
026  pw.print("&amp;Category=$(Category)");
027  pw.print("&amp;Action=Set");
028  pw.print("&amp;Name=$(Name)");
029  pw.println("\"/>");
030  pw.println("</do>");

031  pw.println("<p>");
032  pw.println("Item Name:");
033  pw.println("<input type=\"text\" name=\"Name\"/>");
034  pw.println("</p>");

035  pw.print("<p>Category:");

036  Enumeration enum = userRecord.getCategories();
037  printSelectList(pw, "Category", enum, false, false);

038  pw.println("</p>");
039  pw.println("</card>");
040  pw.println("</wml>");

041  pw.close();
042 }
```

Figure 6.31b The `CategoryItemHandler.handleAdd()` method (Continued).

Next, the `CategoryRecord` is established (lines 010–014). Then, the value of the `Picked` parameter is obtained (line 018) and searched for each category ID (lines 019–027). Any item that is found is marked as picked (line 025). Finally, the `UserRecord` instance is saved, and the user receives a confirmation message (lines 027–028).

The StoresHandler Class

The primary function of the `StoresHandler` class is to process requests associated with the collection of stores belonging to the requesting user. An instance of the `StoresHandler` is invoked by the `Grocery.doPost()` method, in response to having received a request containing the parameter `Type=Store`. Table 6.6 shows a list of query string parameters that are used to invoke the class instance, the source of that type of request, and a description of the action taken.

The handleView( ) Method

The `handleView()` method of the `StoresHandler` class (shown in Figure 6.35b) is used to display the list of `StoreRecord` class instances associated with the user. The first task is to obtain the user's `UserRecord` class instance from the session (lines 006–007). After some standard preliminaries, the cache control is set for forwards and backwards volatility (line 013). The list of stores is displayed by using the `printList-`

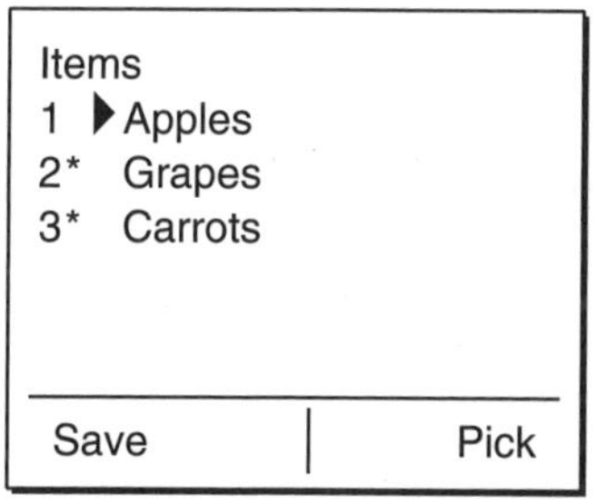

Figure 6.32a A `handleViewItemPicker()` method example.

```
001 protected void handleViewItemPicker(PrintWriter pw,
002   String strCategory,
003   CategoryRecord categoryRecord)
004 {
005   CharArrayWriter caw = new CharArrayWriter(1024);
006   PrintWriter pwOptions = new PrintWriter(caw);

007   String strPicked = "";

008   Enumeration enum = categoryRecord.getItems();

009   while (enum.hasMoreElements())
010   {
011     ItemRecord itemRecord = (ItemRecord) enum.nextElement();

012     if (itemRecord.isPicked())
013     {
014       if (strPicked.length() != 0)
015       {
016         strPicked += ";";
017       }

018       strPicked += itemRecord.getID();
019     }

020     String strItemName = itemRecord.getName();

021     String strItemID = itemRecord.getID();

022     pwOptions.print("<option value=\"" + strItemID);
023     pwOptions.print("\">" + strItemName);
024     pwOptions.println("</option>");
```

(continues)

Figure 6.32b The `handleViewItemPicker()` method.

```
025  }

026  pwOptions.close();

027  pw.println("<onevent type=\"onenterforward\">");
028  pw.println("<refresh>");
029  pw.println("<setvar name=\"Picked\" value=\""+strPicked+"\"/>");
030  pw.println("</refresh>");
031  pw.println("</onevent>");

032  pw.println("<do type=\"accept\" label=\"Save\">");
033  pw.print("<go href=\"");
034  pw.print("?Type=CategoryItem");
035  pw.print("&amp;Action=PickItems");
036  pw.print("&amp;Category=" + strCategory);
037  pw.print("&amp;Picked=$(Picked)");
038  pw.println("\"/>");
039  pw.println("</do>");

040  pw.println("<p><b>Items</b></p>");

041  pw.print("<p><select multiple=\"true\"");
042  pw.println("name=\"Picked\">");

043  pw.write(caw.toCharArray());

044  pw.println("</select></p>");
045 }
```

Figure 6.32b The `handleViewItemPicker()` method (Continued).

`Entries()` method of the `Handler` class. Finally, a single link is added after the stores for adding new stores (lines 018–021).

Figure 6.35a shows an example of the `handleView()` method in action. Once again, we start from the main menu in panel one. The second panel shows the deck rendered by this method when there are no stores. Panel three shows the deck rendered by the `StoresHandler.handleAdd()` method. After the `StoresHandler.handleSet()` method creates the new store, it renders the add confirmation screen shown in panel four. Backing up the history stack, the `handleView()` method is once again invoked via the revalidate cache control, and panel five appears.

The AreasHandler Class

The primary function of the `AreasHandler` class is to process requests associated with the collection of store areas belonging to the requesting user. An instance of the `AreasHandler` is invoked by the `Grocery.doPost()` method in response to having

```
001 public void handlePickItems(HttpServletRequest req,
002  HttpServletResponse resp, HttpSession session)
003  throws ServletException, IOException
004 {
005  String strName = "";

006  UserRecord userRecord =
007    (UserRecord) session.getValue("UserRecord");

008  String strCategory = req.getParameter("Category");
009  CategoryRecord categoryRecord = null;

010  if (userRecord != null)
011  {
012    categoryRecord =
013      userRecord.getCategory(strCategory);
014  }

015  if (categoryRecord != null)
016  {
017    strName = categoryRecord.getName();

018    String strPicked = ";" + req.getParameter("Picked") + ";";

019    Enumeration enum = categoryRecord.getItems();

020    while (enum.hasMoreElements())
021    {
022      ItemRecord itemRecord = (ItemRecord) enum.nextElement();
023      String strItem = ";" + itemRecord.getID() + ";";

024      boolean bPicked = (strPicked.indexOf(strItem) != -1);

025      itemRecord.setPicked(bPicked);
026    }

027    save(userRecord);
028    printConfirmation(resp, "<b>Category saved!</b>");
029  }
030  else
031  {
032    resp.sendError(HttpServletResponse.SC_BAD_REQUEST);
033  }
034 }
```

Figure 6.33 The `CategoryItemHandler.handlePickItems()` method.

Table 6.6 StoresHandler Query Strings

QUERY STRING	DESCRIPTION
?Type=Store&Action=View	Displays a list of stores associated with a user
?Type=Store&Action=Add	Displays a form for entering a new store
?Type=Store&Action=Set&Name=<name>	Establishes a new store

```
package book.chap06.Grocery;

public class StoresHandler extends Handler
{
  public StoresHandler()
    throws NoSuchMethodException;
  public void run(HttpServletRequest req,
    HttpServletResponse resp,
    HttpSession session)
    throws ServletException, IOException;
  public void handleView(HttpServletRequest req,
    HttpServletResponse resp,
    HttpSession session)
    throws ServletException, IOException;
  public void handleAdd(HttpServletRequest req,
    HttpServletResponse resp,
    HttpSession session)
    throws ServletException, IOException;
  public void handleSet(HttpServletRequest req,
    HttpServletResponse resp,
    HttpSession session)
    throws ServletException, IOException;
}
```

Figure 6.34 The StoresHandler prototype.

received a request containing the parameter Type=Area. Table 6.7 shows a table of query string parameters that are used to invoke the class instance, the source of that type of request, and a description of the action taken.

The handleView( ) Method

The handleView() method of the AreaHandler method (shown in Figure 6.37b) is one of the more complex handler methods. Due to the atypical way in which the list of areas is processed, all of the work is done within this method—rather than depending on methods in the Handler class.

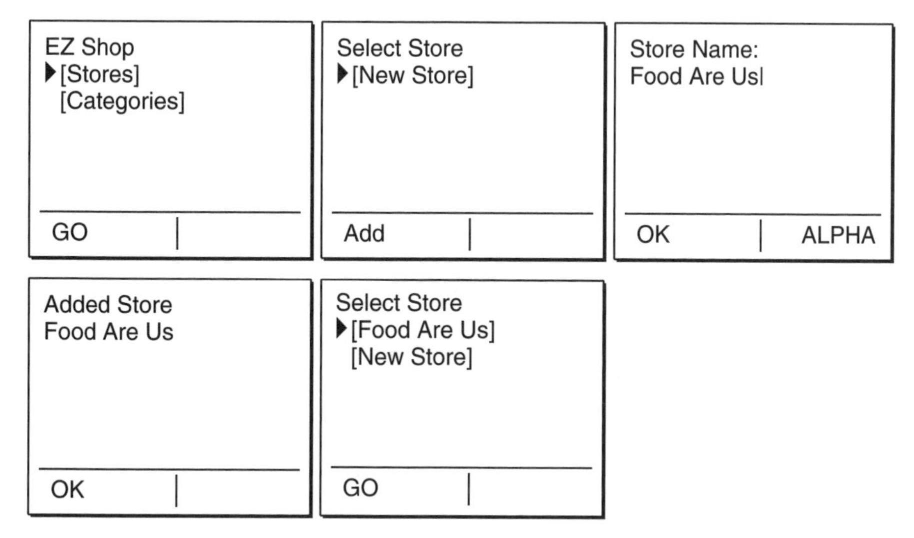

Figure 6.35a A `StoresHandler.handleView()` example.

```
001 public void handleView(HttpServletRequest req,
002  HttpServletResponse resp,
003  HttpSession session)
004  throws ServletException, IOException
005 {
006  UserRecord userRecord =
007    (UserRecord) session.getValue("UserRecord");

008  if (userRecord != null)
009  {
010    PrintWriter pw = resp.getWriter();

011    printXML(pw);

012    pw.println("<wml>");
013    printHeader(pw, true, true, true);

014    pw.println("<card>");

015    pw.println("<p><b>Select Store</b></p>");

016    Enumeration enum = userRecord.getStores();
017    printListEntries(enum, pw, "Area", "Store");

018    pw.print("<p><i><a title=\"Add\" href=\"");
```

(continues)

Figure 6.35b The `StoresHandler.handleView()` method.

```
019     pw.print("?Type=Store");
020     pw.print("&amp;Action=Add");
021     pw.println("\">New Store</a></i></p>");

022     pw.println("</card>");
023     pw.println("</wml>");
024     pw.close();
025   }
026   else
027   {
028     resp.sendError(HttpServletResponse.SC_BAD_REQUEST);
029   }
030 }
```

Figure 6.35b The `StoresHandler.handleView()` method (Continued).

Table 6.7 `AreasHandler` Query Strings

QUERY STRING	DESCRIPTION
?Type=Area&Action =View&Store=<id>	Displays the areas within a store
?Type=Area&Action =Add&Store=<id>	Displays a form for adding a new area to a store
?Type=Area&Action=Set&Store =<id>&Area=<id>&Name=<name>	Establishes a new area
?Type=Area&Action =Unassigned&Store=<id>	Views a list of unassigned items

Because this method is used to view the areas within one of the user's stores, we need to have references to the `UserRecord` instance (lines 007–008) and to the `StoreRecord` instance (lines 009–013). Provided that we can get these references, the next order of business is to perform the basic rendering of the early deck and card elements (lines 016–022). Because information on this deck is volatile in both forward and backward directions, no-cache and revalidate are both enabled (line 020).

Because a list of store areas is to be displayed, the next step is to iterate through the enumeration of store areas (lines 024–053). When a user is shopping, it will be beneficial for them to see what current areas have picked items. Therefore, we would like to put an asterisk in front of each area that has picked items. To achieve this goal, we must iterate through the items in each area until one picked item is found (lines 029–037).

Immediately after the list of areas, we would like to put a link to the `handleUnassigned()` method (lines 065–072). This link is only needed, however, if there are picked, unassigned items associated with the store. Therefore, the unassigned items of the store are iterated until a picked, unassigned item is found (lines 055–064). Finally, a link to the `addHandler()` is appended to the list (lines 073–077).

```
package book.chap06.Grocery;

public class AreasHandler extends Handler
{
  public AreasHandler()
    throws NoSuchMethodException;
  public void run(HttpServletRequest req,
    HttpServletResponse resp,
    HttpSession session)
    throws ServletException, IOException;
  public void handleView(HttpServletRequest req,
    HttpServletResponse resp,
    HttpSession session)
    throws ServletException, IOException;
  public void handleUnassigned(HttpServletRequest req,
    HttpServletResponse resp,
    HttpSession session)
    throws ServletException, IOException;
  protected void printNoAreaNotification(HttpServletResponse resp)
    throws IOException;
  public void handleAdd(HttpServletRequest req,
    HttpServletResponse resp,
    HttpSession session)
    throws ServletException, IOException;
  public void handleSet(HttpServletRequest req,
    HttpServletResponse resp,
    HttpSession session)
    throws ServletException, IOException;
}
```

Figure 6.36 The `AreasHandler` prototype.

NOTE

When a new store is added, no attempt is made to add existing items to the store. You could implement this feature by traversing the `RecordList` containing the `ItemRecord` references associated with a user (in the `UserRecord`). I did not choose to implement this function, however, because it would result in a lot of work for the user to clear out his or her unassigned item list. Each new item added to a category, however, is added to each existing store's unassigned items.

Figure 6.37a shows an example screen shot of a deck generated by the `handleView()` method. This example shows the area list for a store. Items in the produce and aisle 2 areas are picked. No items are picked in the aisle 1 area. Due to the fact that the unassigned link is available, we know that there are picked, unassigned items associated with the store.

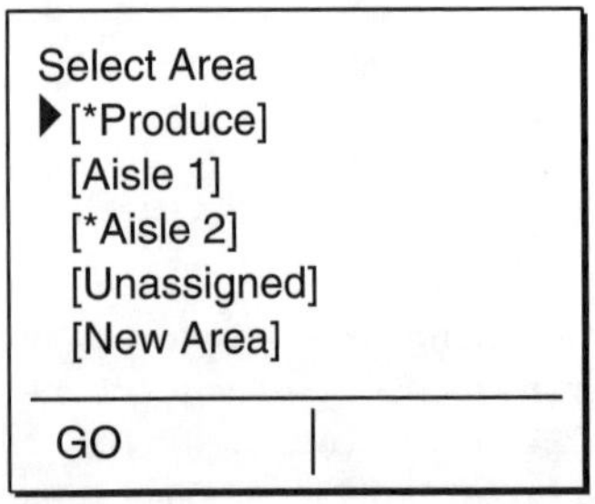

```
Select Area
▶[*Produce]
 [Aisle 1]
 [*Aisle 2]
 [Unassigned]
 [New Area]
---------------
 GO
```

Figure 6.37a An `AreasHandler.handleView()` example.

```
001 public void handleView(HttpServletRequest req,
002  HttpServletResponse resp,
003  HttpSession session)
004  throws ServletException, IOException
005 {
006  String strStoreID = req.getParameter("Store");

007  UserRecord userRecord =
008    (UserRecord) session.getValue("UserRecord");

009  StoreRecord storeRecord = null;
010  if ((userRecord != null) && (strStoreID != null))
011  {
012    storeRecord = userRecord.getStore(strStoreID);
013  }

014  if (storeRecord != null)
015  {
016    String strStoreName = storeRecord.getName();
017    PrintWriter pw = resp.getWriter();
018    printXML(pw);

019    pw.println("<wml>");

020    printHeader(pw, true, true, true);

021    pw.println("<card>");

022    pw.println("<p><b>Select Area</b></p>");

023    Enumeration enum = storeRecord.getAreas();
024    while (enum.hasMoreElements())
025    {
```

Figure 6.37b The `AreasHandler.handleView()` method.

```
026       AreaRecord areaRecord = (AreaRecord) enum.nextElement();

027       Enumeration enum2 = areaRecord.getItems();
028       boolean bFlagIt = false;

029       while (enum2.hasMoreElements())
030       {
031         ItemRecord itemRecord =
              (ItemRecord) enum2.nextElement();

032         if (itemRecord.isPicked())
033         {
034           bFlagIt = true;
035           break;
036         }
037       }

038       String strName = areaRecord.getName();
039       String strID = areaRecord.getID();

040       pw.print("<p mode=\"nowrap\">");
041       pw.print("<a title=\"GO\" href=\"?Type=AreaItem");
042       pw.print("&amp;Action=View");
043       pw.print("&amp;Area=" + strID);
044       pw.print("&amp;Store=" + strStoreID);
045       pw.print("\">");

046       if (bFlagIt)
047       {
048         pw.print("*");
049       }

050       pw.print("<![CDATA[" + strName + "]]>");
051       pw.print("</a>");
052       pw.println("</p>");
053     }

054     boolean bUnassigned = false;

055     enum = storeRecord.getUnassignedItems();
056     while (enum.hasMoreElements())
057     {
058      ItemRecord item = (ItemRecord) enum.nextElement();

059      if (item.isPicked())
```

(continues)

Figure 6.37b The `AreasHandler.handleView()` method.

```
060         {
061            bUnassigned = true;
062            break;
063          }
064       }

065       if (bUnassigned)
066       {
067         pw.print("<p><a title=\"GO\" href=\"");
068         pw.print("?Type=Area");
069         pw.print("&amp;Action=Unassigned");
070         pw.print("&amp;Store=" + strStoreID);
071         pw.println("\">Unassigned</a></p>");
072       }

073       pw.print("<p><i><a title=\"Add\" href=\"");
074       pw.print("?Type=Area");
075       pw.print("&amp;Action=Add");
076       pw.print("&amp;Store=" + strStoreID);
077       pw.println("\">New Area</a></i></p>");

078       pw.println("</card>");

079       pw.println("</wml>");
080       pw.close();
081     }
082     else
083     {
084       resp.sendError(HttpServletResponse.SC_BAD_REQUEST);
085     }
086   }
```

Figure 6.37b The `AreasHandler.handleView()` method (Continued).

The handleUnassigned( ) Method

The `handleUnassigned()` method uses techniques that we have seen in prior methods to assign unassigned items associated with a store to one of the stores categories. The source code for this method is shown in Figure 6.38b. In order for this method to perform its task, it needs to have the `UserRecord` and `StoreRecord` instance (lines 006–013). If there are no areas available for assignment, the `printNoAreaNotification()` method is called (lines 020–023). Next, we set up the *accept* soft key in order to call the `handleAssign()` method of the `AreaItemHandler` class with the item (in variable `$(Item)`) and area (in variable `$(Area)`) selected (lines 032–037). Finally, the `printSelectList()` method is called to render a `<select>` element's picked items (lines 041–042) and for all areas (lines 044–045).

Figure 6.38a shows an example of the `handleUnassigned()` method. Panel one shows the selection of the *unassigned* link that invokes this method. Panel two shows the first

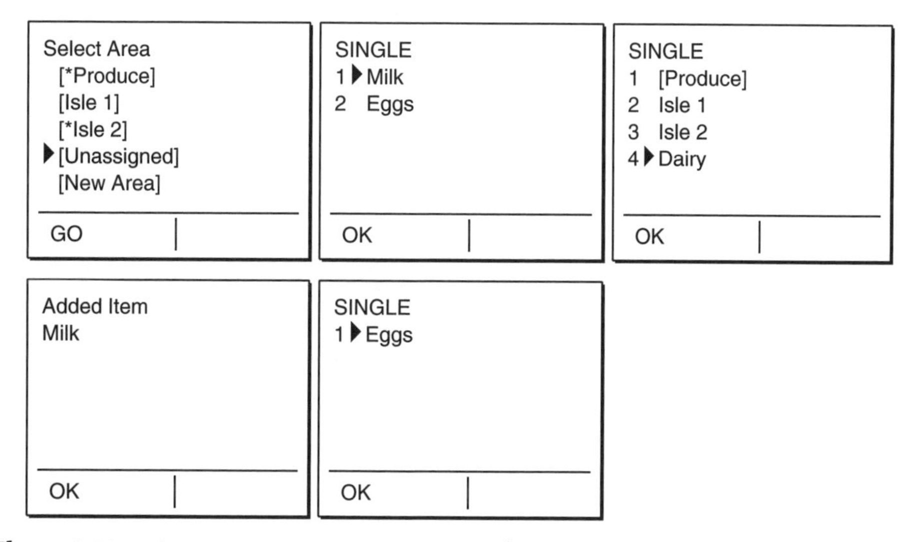

Figure 6.38a A `handleUnassigned()` example.

```
001 public void handleUnassigned(HttpServletRequest req,
002  HttpServletResponse resp,
003  HttpSession session)
004  throws ServletException, IOException
005 {
006  String strStoreID = req.getParameter("Store");

007  UserRecord userRecord =
008    (UserRecord) session.getValue("UserRecord");

009  StoreRecord storeRecord = null;
010  if ((userRecord != null) && (strStoreID != null))
011  {
012    storeRecord = userRecord.getStore(strStoreID);
013  }

014  if (storeRecord != null)
015  {
016    if (storeRecord.getUnassignedItemCount() == 0)
017    {
018      printConfirmation(resp, "<b>All items assigned!</b>");
019    }
020    else if (storeRecord.getAreaCount() == 0)
021    {
022      printNoAreaNotification(resp);
```

(continues)

Figure 6.38b The `handleUnassigned()` method.

```
023     }

024     else
025     {
026       PrintWriter pw = resp.getWriter();

027       printXML(pw);

028       pw.println("<wml>");
029       printHeader(pw, true, true, true);
030       pw.println("<card>");
031       pw.println("<do type=\"accept\">");

032       pw.print("<go href=\"?Type=AreaItem&amp;");
033       pw.print("Action=Assign");
034       pw.print("&amp;Store=" + strStoreID);
035       pw.print("&amp;Area=$(Area)");
036       pw.print("&amp;Item=$(Item)");
037       pw.println("\"/>");

038       pw.println("</do>");

039       pw.println("<p>");
040       pw.println("<b>Select Item</b>");
041       Enumeration enum = storeRecord.getUnassignedItems();
042       printSelectList(pw, "Item", enum, true, false);

043       pw.println("<b>Select Area</b>");
044       enum = storeRecord.getAreas();
045       printSelectList(pw, "Area", enum, false, false);

046       pw.println("</p>");
047       pw.println("</card>");

048       pw.println("</wml>");
049       pw.close();
050   }
051   }
052   else
053   {
054     resp.sendError(HttpServletResponse.SC_BAD_REQUEST);
055   }
056 }
```

Figure 6.38b The `handleUnassigned()` method (Continued).

`<select>` element (for selecting the unassigned item). Panel three shows the second `<select>` element (for selecting an area assigned to the item). When the *accept* soft key is pressed on the third panel, the handleAssign() method of the AreaItemHandler class is invoked. Once handleAssign() moves the selected item to the selected area, it renders the add confirmation. The revalidated deck generated by this method is then showed in panel five.

The printNoAreaNotification( ) Method

The printNoAreaNotification() method is used to display a message to the user when he or she enters the unassigned area without having areas available for assignment. The source code for this method is shown in Figure 6.39.

WARNING

Although the message generated by this method is not volatile, it is invoked via a URL that is shared with the volatile deck rendered by the handleUnassigned() method. Therefore, it must be set for no-cache (line 007). This rule is subtle, yet important.

The AreaItemHandler Class

The primary function of the AreaItemHandler class (shown in Figure 6.40) is to process requests associated with a specific area or the area's items. An instance of the AreaItemHandler is invoked by the Grocery.doPost() method in response to having received a request containing the parameter Type=AreaItem. Table 6.8 shows a table of query string parameters that are used to invoke the class instance, the source of that type of request, and a description of the action taken.

```
001 protected void printNoAreaNotification(HttpServletResponse resp)
002  throws IOException
003 {
004  PrintWriter pw = resp.getWriter();
005  printXML(pw);

006  pw.println("<wml>");

007  printHeader(pw, true, true, false);

008  pw.println("<card>");
009  pw.println("<p>In order to assign unassigned items, you");
010  pw.println("must have at least one area. Back up and use");
011  pw.println("<i>New Area</i> to add an area.");

012  pw.println("</p></card>");
013  pw.println("</wml>");

014  pw.close();
015 }
```

Figure 6.39 The AreasHandler.printNoAreaNotifcation() method.

```
package book.chap06.Grocery;

public class AreaItemHandler extends Handler
{
  public AreaItemHandler()
    throws NoSuchMethodException;
  public void run(HttpServletRequest req,
    HttpServletResponse resp,
    HttpSession session)
    throws ServletException, IOException;
  public void handleView(HttpServletRequest req,
    HttpServletResponse resp,
    HttpSession session)
    throws ServletException, IOException;
  protected void printNoItemNotification(HttpServletResponse resp)
    throws IOException;
  public void handleAssign(HttpServletRequest req,
    HttpServletResponse resp,
    HttpSession session)
    throws ServletException, IOException;
  public void handleCheck(HttpServletRequest req,
    HttpServletResponse resp,
    HttpSession session)
    throws ServletException, IOException;
}
```

Figure 6.40 `AreaItemHandler` prototype.

Table 6.8 `AreaItemHandler` Query Strings

QUERY STRING	DESCRIPTION
`?Type=AreaItem&Action=View&Store=<id>&Area=<id>`	Displays the items within an area
`?Type=AreaItem&Action=Assign&Store=<id>&Area=<id>&Item=<id>`	Assigns an item to the area
`?Type=AreaItem&Action=Check&Store=<id>&Area=<id>&Picked=<ids>`	Displays a check-off list

The handleView( ) Method

The `handleView()` method (shown in Figure 6.41b) is similar to other `handleView()` methods that we have seen; therefore, we will not discuss this topic in depth. Of primary interest in this implementation is that only picked items in the area are shown in a multi-item `<select>` element (line 051). They are displayed as non-selected, however, because the `$(Picked)` variable is initialized to empty (lines 036–040). The *accept* soft key is configured to invoke the `handleCheck()` method (lines 041–047).

Figure 6.41a shows an example of the deck rendered by `handleView()`. In this example, the user has checked off two items. Therefore, if the user saved and re-entered this list, the list would only contain the picked items (it would not contain those that have been checked).

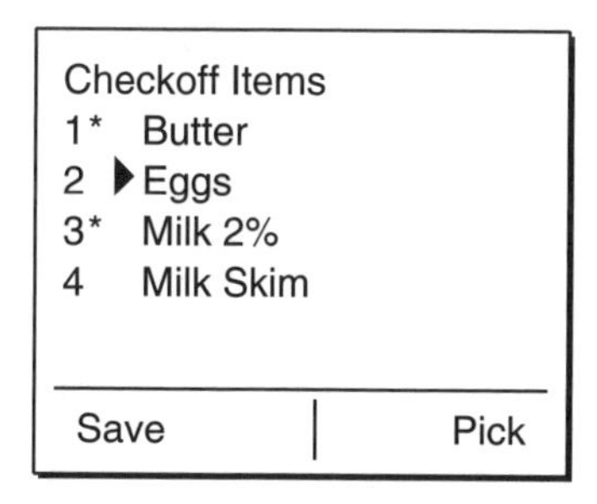

Figure 6.41a The `AreaItemHandler.handleView()` method.

```
001 public void handleView (HttpServletRequest req,
002  HttpServletResponse resp,
003  HttpSession session)
004  throws ServletException, IOException
005 {
006  String strStoreID = req.getParameter("Store");
007  String strAreaID = req.getParameter("Area");

008  UserRecord userRecord =
009   (UserRecord) session.getValue("UserRecord");

010  StoreRecord storeRecord = null;

011  if (userRecord != null)
012  {
013    storeRecord = userRecord.getStore(strStoreID);
014  }

015  AreaRecord areaRecord = null;

016  if (storeRecord != null)
017  {
018    areaRecord = storeRecord.getArea(strAreaID);
019  }

020  if (areaRecord != null)
021  {
022    if (areaRecord.getItemCount() == 0)
```

(continues)

Figure 6.41b The `AreaItemHandler.handleView()` method.

```
023    {
024      printNoItemNotification(resp);
025    }
026    else
027    {
028      PrintWriter pw = resp.getWriter();
029      printXML(pw);
030      pw.println("<wml>");
031      printHeader(pw, true, true, false);
032      pw.println("<card>");

033      pw.println("<onevent type=\"onenterbackward\">");
034      pw.println("<prev/>");
035      pw.println("</onevent>");

036      pw.println("<onevent type=\"onenterforward\">");
037      pw.println("<refresh>");
038      pw.println("<setvar name=\"Picked\" value=\"\"/>");
039      pw.println("</refresh>");
040      pw.println("</onevent>");

041      pw.println("<do type=\"accept\" label=\"Save\">");
042      pw.print("<go href=\"?Type=AreaItem&amp;Action=Check");
043      pw.print("&amp;Store=" + strStoreID);
044      pw.print("&amp;Area=" + strAreaID);
045      pw.print("&amp;Picked=$(Picked)");
046      pw.println("\"/>");
047      pw.println("</do>");

048      pw.println("<p><b>Checkoff Items</b></p>");

049      pw.println("<p>");

050      Enumeration enum = areaRecord.getItems();
051      printSelectList(pw, "Picked", enum, true, true);

052      pw.println("</p>");

053      pw.println("</card>");
054      pw.println("</wml>");
055      pw.close();
056    }
057  }
058  else
059  {
060   resp.sendError(HttpServletResponse.SC_BAD_REQUEST);
061  }
062 }
```

Figure 6.41b The `AreaItemHandler.handleView()` method (Continued).

The printNoItemNotification( ) Method

The `printNoItemNotification()` method (shown in Figure 6.42) is used to display a message to the user indicating that a store area is empty.

TIP

Due to the way in which we navigate through WML decks, this situation is not unusual. While I could have simply excluded empty areas in the `handleView()` method of the `AreasHandler` class, I did not want to perform this action—because the user might add the area again if he or she did not see that it already existed (from a previous session, perhaps).

The handleAssign( ) Method

The `handleAssign()` method is invoked in response to assigning an unassigned item to a store area via the WML deck rendered by the `handleUnassigned()` method. Figure 6.43 shows the implementation of this method.

The handleCheck( ) Method

The `handleCheck()` method is used to process the check-off list generated by the `AreaItemHandler.handleView()` method. In order to perform this task, the method obtains the value of the `$(Picked)` variable (line 008). Once the `AreaRecord` instance is established (lines 009–020), the items of the area are iterated (lines 024–033). As each item is obtained, it is marked as not picked if its ID is found in the `$(Picked)` variable. Refer to Figure 6.44.

```
001 protected void printNoItemNotification(HttpServletResponse resp)
002  throws IOException
003 {
004  PrintWriter pw = resp.getWriter();
005  printXML(pw);

006  pw.println("<wml>");
007  printHeader(pw, true, true, false);

008  pw.println("<card>");
009  pw.println("<p>No items have been assigned to this store");
010  pw.println("area. Add new items to a category then");
011  pw.println("assign those unassigned items to an area.");

012  pw.println("</p></card>");

013  pw.println("</wml>");

014  pw.close();
015 }
```

Figure 6.42 The `AreaItemHandler.printNoItemNotifcation()` method.

```
001 public void handleAssign(HttpServletRequest req,
002  HttpServletResponse resp,
003  HttpSession session)
004  throws ServletException, IOException
005 {
006  String strStoreID = req.getParameter("Store");
007  String strAreaID = req.getParameter("Area");
008  String strItemID = req.getParameter("Item");

009  UserRecord userRecord =
010    (UserRecord) session.getValue("UserRecord");

011  StoreRecord storeRecord = null;

012  if (userRecord != null)
013  {
014    storeRecord = userRecord.getStore(strStoreID);
015  }

016  AreaRecord areaRecord = null;

017  if (storeRecord != null)
018  {
019    areaRecord = storeRecord.getArea(strAreaID);
020  }

021  ItemRecord itemRecord = null;

022  if (areaRecord != null)
023  {
024    itemRecord = userRecord.getItem(strItemID);
025  }

026  if (itemRecord != null)
027  {
028    areaRecord.addItem(itemRecord);
029    storeRecord.removeUnassignedItem(strItemID);
030    confirmAddition(resp, "Item", itemRecord.getName());
031  }
032  else
033  {
034    resp.sendError(HttpServletResponse.SC_BAD_REQUEST);
035  }
036 }
```

Figure 6.43 The `AreaItemHandler.handleAssign()` method.

```
001 public void handleCheck(HttpServletRequest req,
002  HttpServletResponse resp,
003  HttpSession session)
004  throws ServletException, IOException
005 {
006  String strStoreID = req.getParameter("Store");
007  String strAreaID = req.getParameter("Area");
008  String strPicked = ";" + req.getParameter("Picked") + ";";

009  UserRecord userRecord =
010    (UserRecord) session.getValue("UserRecord");

011  StoreRecord storeRecord = null;

012  if (userRecord != null)
013  {
014    storeRecord = userRecord.getStore(strStoreID);
015  }

016  AreaRecord areaRecord = null;

017  if (storeRecord != null)
018  {
019      areaRecord = storeRecord.getArea(strAreaID);
020  }

021  if (areaRecord != null)
022  {
023    Enumeration enum = areaRecord.getItems();

024    while (enum.hasMoreElements())
025    {
026      ItemRecord itemRecord = (ItemRecord) enum.nextElement();

027      String strID = itemRecord.getID();

028      if (itemRecord.isPicked() &&
029        (-1 != strPicked.indexOf(";" + strID + ";")))
030      {
031        itemRecord.setPicked(false);
032      }
033    }

034    printConfirmation(resp, "<b>Area list saved.</b>");
035  }
036  else
```

Figure 6.44 The `AreaItemHandler.handleCheck()` method.

```
037  {
038     resp.sendError(HttpServletResponse.SC_BAD_REQUEST);
039  }
040 }
```

Figure 6.44 The `AreaItemHandler.handleCheck()` method (Continued).

Ramifications

Obviously, this application is a rather simple example application. There are a couple of modifications that could be made to make the application commercial grade. The following list describes these modifications:

- Users should have a username and password. If an existing user attempt to access his or her account via a new client, the user should be able to log on to the existing account; otherwise, a new user should be able to select an option to create a new account.
- Rather than using a flat-file data-storage mechanism, use of a Java Database Connection (JDBC) database (such a MySQL) would be more scaleable.
- Features should be added to remove, rename, and reorder the stores, categories, and items.
- Being able to transfer items between categories and store areas would also be helpful.

Because many of these operations go beyond the types of activities that a user might want to perform on a WAP browser, a complementary HTML interface into a servlet could enable a user to access the extended features list via a standard Web browser.

Conclusion

In this chapter, we saw how a WAP application can be realized by using WML and JSDK Servlet technology. By now, the power and flexibility of this technology should be apparent. In the next chapter, we will look at another technology associated with WAP: Push.

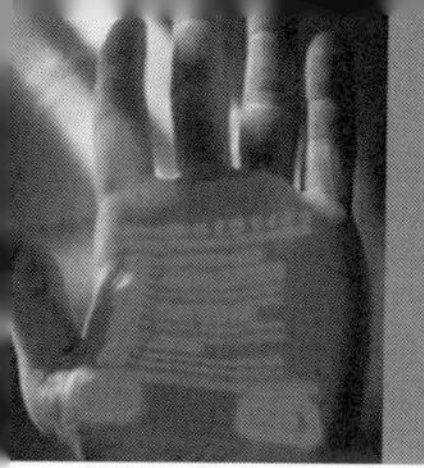

CHAPTER 7

Push Technology

In the previous chapter, everything that we learned about *Wireless Markup Language* (WML) and the *Java Servlet Development Kit* (JSDK) Servlets came together in a simple, real-world application. By now, you must find it easy to see why I referred to this technology as a paradigm shift. By using these capabilities, users can access simple content and applications by using small, hand-held clients that are untethered. There is another element of *Wireless Application Protocol* (WAP) technology that helps the user extend his or her desktop, however. In this chapter, we will examine Push technology and how it will simplify and increase the power of the WAP browser.

Introduction

So far, we have seen that WML provides the webmaster with the capability to provide Web content and applications that are suitable for processing by a WAP client. Because we have dealt so far with the content that a servlet engine requests and delivers, we have limited WAP to the synchronous request and response experience made popular with *Hypertext Markup Language* (HTML) browsers. This capability is known as Pull technology—so named because the user's client device requisitions a particular network resource that is pulled from the server. The WAP Push standard describes a capability that enables unsolicited, asynchronous messages to be sent by a server to a browser, however. Therefore, Pull technology is so named also because the data is pushed to the client.

Push Hardware Architecture

In the WAP arena, Push hardware architecture consists of three elements: a *Push Initiator* (PI), a *Push Proxy Gateway* (PPG), and a client. The client is obviously the same client that is used to request and display WML content. The Push gateway tends to be the WAP gateway used to bridge the Internet to a bearer network. The PI is any system that has the capability to source Push messages through a Push gateway for a WAP client. Figure 7.1 shows the WAP Push hardware architecture.

The Push gateway serves two purposes in the architecture. First, it translates a standard Push message delivered over the Internet from the initiator to the client via an appropriate protocol over the bearer network. Second, it communicates the outcome of a Push request to the originator.

Push Access Protocol (PAP)

From a protocol perspective, the *Push Access Protocol* (PAP) is used to create and track the progress of Push messages. By using PAP, you can send several types of request and response elements, such as the following: `<push-message>`, `<push-response>`, `<cancel-message>`, `<cancel-response>`, `<resultnotification-message>`, `<resultnotifcation-response>`, `<statusquery-message>`, `<statusquery-response>`, and `<badmessage-response>`. In the subsections to come, we will examine these requests in more detail. Figure 7.2 shows the `<pap>` element of the PAP protocol Document Template Definition (DTD).

The <push-message> Element

The `<push-message>` element is the basic push mechanism in the PAP. By using this request, you can send a message to one or more clients. When using this request, the PI might specify the following attributes: `push-id`, `deliver-before-timestamp`, and `deliver-after-timestamp`. The `push-id` is a unique identifier defined by the PI that you can use to cancel a request in the future. The `deliver-before-timestamp` attribute is used to specify a time after which the PPG must not deliver the message (an expiration time). The `deliver-after-timestamp` attribute is used to establish a time after which the PPG can deliver the message.

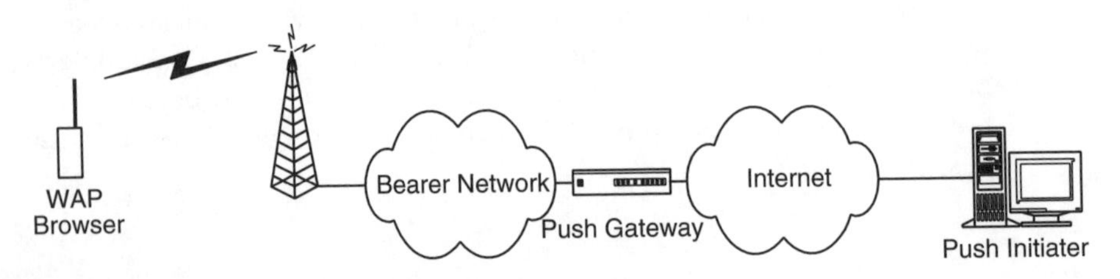

Figure 7.1 Push hardware architecture.

```
<!ELEMENT pap                       ( push-message
                                    | push-response
                                    | cancel-message
                                    | cancel-response
                                    | resultnotification-message
                                    | resultnotification-response
                                    | statusquery-message
                                    | statusquery-response
                                    | ccq-message
                                    | ccq-response
                                    | badmessage-response) >
<!ATTLIST pap
          product-name              CDATA            #IMPLIED
>
```

Figure 7.2 The <pap> element.

NOTE

Besides serving the function of a gateway between the Internet and the bearer network, the PPG provides another import function with regard to asynchronous message delivery. Because at any given time the WAP client might be turned off or out of range, the PPG must hold and attempt to deliver a message until it has confirmation that a message has been delivered. The `deliver-before-timestamp` attribute becomes important, because a message might become obsolete if it is not delivered within a particular time.

The PPG also has the capability to report the status of a message to the PI when the message utilizes two additional attributes: `ppg-notify-requested-to` and `progress-notes-requested`. The `ppg-notify-requested-to` attribute is used to specify an address (typically, a *Uniform Resource Locator* [URL]) used for notification of results. When the value of the `progress-notes-requested` attribute has a value of *true*, progress notes are sent by the PPG to the `ppg-notify-request-to` address.

NOTE

Using the notification capabilities of the PPG, a PI can monitor the delivery of a message. When the data in a message is time sensitive (for example, a meeting reminder), the `deliver-before-timestamp` is the preferred method for expiring a message. When a message is data sensitive, however (for example, a stock trade recommendation), the notification technique is favored.

Figure 7.3 shows the code fragment that defines the <push-message> element.

The <push-message> element contains one or more <address> elements. Each address element is used to define the address of a single recipient that is defined via the

```
<!ELEMENT push-message ( address+, quality-of-service? ) >
<!ATTLIST push-message
        push-id                      CDATA                #REQUIRED
        deliver-before-timestamp     %Datetime;           #IMPLIED
        deliver-after-timestamp      %Datetime;           #IMPLIED
        source-reference             CDATA                #IMPLIED
        ppg-notify-requested-to      CDATA                #IMPLIED
        progress-notes-requested     ( true | false )     "false"
>
```

Figure 7.3 The `<push-message>` element.

```
<!ELEMENT address EMPTY >
<!ATTLIST address
        address-value          CDATA                #REQUIRED
>
```

Figure 7.4 The `<address>` element.

`address-value` attribute. Figure 7.4 shows a DTD fragment that defines the `<address>` element.

NOTE

The address of the client is largely dependent on the PPG implementation and the bearer network. Therefore, you will need to understand the addressing conventions that are in use for your particular application.

The `<push-message>` element can also contain one optional `<quality-of-service>` element (see Figure 7.5). This element enables the PI to convey its desires for the quality of the message-delivery service offered by the PPG. Six attributes are used to communicate these preferences: `priority`, `delivery-method`, `network`, `network-required`, `bearer`, and `bearer-required`.

The priority of a message (*low*, *medium*, or *high*) is established via the `priority` attribute. The meaning of this attribute and its implementation are PPG dependent. In fact, PPGs are not required to implement this attribute at all. The PPG could use this value to reorder a long queue of messages waiting to be sent.

TIP

Avoid using the high-priority setting for all messages. If everyone used this setting, it would make the priority attribute meaningless. If you use medium or low priority for routine messages and high priority for important (time-critical) messages, you will benefit from this scheme—provided that your PPG implements the `priority` attribute.

The <delivery-method> attribute takes one of four values: *confirmed*, *preferconfirmed*, *unconfirmed*, or the default *notspecified*. If the value is *confirmed*, the PPG is required to confirm that a message has been delivered. If the value is *preferconfirmed*, the PPG is required to attempt *confirmed* delivery of the message but can use another method if that method does not succeed. If the value is *unconfirmed*, the PPG must deliver the message by using a *non-confirmed* method delivery. A value of *notspecified* indicates that the PI does not care what method is used and that the PPG can choose the method.

NOTE

If you application is sending an important message, such as a confirmation of the completion of a financial transaction, you should use the *confirmed* method. If your application periodically sends routine information to subscribers, however (for example, alternate traffic routes), you should use the *unconfirmed* method to avoid needless processing and bandwidth utilization.

If you would like to specify or suggest a network for delivering the message, you can specify the network and network-required attributes. The network attribute suggests a network type. The network-required attribute makes your suggestion a requirement.

Additionally, you can specify or suggest a bearer for delivering the message via the bearer and bearer-required attributes.

The <push-response> Element

The <push-response> is the basic response mechanism. By using this response, the PPG can notify the PI about the outcome of a previously submitted <push-message>.

```
<!ELEMENT quality-of-service EMPTY >
<!ATTLIST quality-of-service
          priority              ( high
                                | medium | low )        "medium"
          delivery-method       ( confirmed
                                | preferconfirmed
                                | unconfirmed
                                | notspecified)         "notspecified"
          network               CDATA                   #IMPLIED
          network-required      ( true | false )        "false"
          bearer                CDATA                   #IMPLIED
          bearer-required       ( true | false )        "false"
>
```

Figure 7.5 The <quality-of-service> element.

The following attributes are used in conjunction with the response: push-id, sender-address, sender-name, and reply-time. The push-id is the unique identifier originally sent in the push-message. The sender-address attribute specifies the URL of the PPG. The sender-name attribute specifies the readable name of the PPG (useful for debugging purposes). The reply-time is the time when the PPG generated the push-response. Figure 7.6 defines the <push-response> element.

The <push-response> can contain one or more optional <progress-note> elements. This element contains three attributes: stage, note, and time. The stage attribute contains a PPG implementation-specific code or text message that indicates the current stage of a message's delivery. The note attribute describes the outcome of the stage. The time attribute specifies the time that the stage completed. Figure 7.7 shows the DTD fragment that defines the <progress-note> element.

NOTE

A PPG might report more than one stage within a single <push-response>. The response can therefore contain more than one <progress-note>.

The <push-response> contains one <response-result> element. This element contains the immediate element status of a <push-message>. The push-response element contains two attributes: code and desc. The code attribute indicates the status code of a request (see Table 7.1), while the desc attribute contains a readable text

```
<!ELEMENT push-response ( progress-note*, response-result ) >
<!ATTLIST push-response
          push-id              CDATA              #REQUIRED
          sender-address       CDATA              #IMPLIED
          sender-name          CDATA              #IMPLIED
          reply-time           %Datetime;         #IMPLIED
>
```

Figure 7.6 The <push-response> element.

```
<!ELEMENT progress-note EMPTY >
<!ATTLIST progress-note
          stage                CDATA              #REQUIRED
          note                 CDATA              #IMPLIED
          time                 %Datetime;         #IMPLIED
>
```

Figure 7.7 The <progress-note> element.

Table 7.1 Push Result Codes

CODE	MEANING
1000	OK
1001	Accepted for Processing
2000	Bad Request
2001	Forbidden
2002	Address Error
2003	Address Not Found
2004	Push ID Not Found
2005	Capabilities Mismatch
2006	Required Capabilities Not Supported
2007	Duplicate Push ID
3000	Internal Server Error
3001	Not Implemented
3002	Version Not Supported
3003	Not Possible
3004	Capability Matching Not Supported
3005	Multiple Addresses Not Supported
3006	Transformation Failure
3007	Specified Delivery Method Not Possible
3008	Capabilities Not Available
3009	Required Network Not Available
3010	Required Bearer Not Available
4000	Service Failure
4001	Service Unavailable
5xxx	Mobile Client Aborted

description of the code. Figure 7.8 shows the DTD fragment that defines the `<response-result>` element. Table 7.1 shows the status codes that can be returned.

The <cancel-message> Element

The `<cancel-message>` request is used to cancel a previously submitted `<push-message>` request. This message contains the `push-id` of the original message as its only attribute. If the original `push-message` was sent to more than one recipient, one

```
<!ELEMENT response-result EMPTY >
<!ATTLIST response-result
        code                    CDATA           #REQUIRED
        desc                    CDATA           #IMPLIED
>
```

Figure 7.8 The <response-result> element.

```
<!ELEMENT cancel-message ( address* ) >
<!ATTLIST cancel-message
        push-id                 CDATA           #REQUIRED
>
```

Figure 7.9 The <cancel-message> element.

```
<!ELEMENT cancel-response ( cancel-result+ ) >
<!ATTLIST cancel-response
        push-id                 CDATA           #REQUIRED
>
```

Figure 7.10 The <cancel-response> element.

or more optional address elements (refer to Figure 7.4) can be specified within the content of the <cancel-message> request. In this case, a pending message will only be canceled for the specified clients. If no <address> elements are specified, however, the pending request will be canceled for all pending recipients. Figure 7.9 shows the DTD fragment that defines the <cancel-message> element.

The <cancel-response> Element

The <cancel-response> is sent by the PPG to the PI in reaction to receiving a <cancel-message> request. The ID specified on the <cancel-message> request is included as the only attribute: push-id. One or more <cancel-result> elements are included in the content of the element. Figure 7.10 shows the DTD fragment that defines the <cancel-response> element.

The <cancel-result> element is sent within the content of a <cancel-response> element. This element provides a means for describing the results of a cancel operation for a particular group of addresses. If the original <push-message> addressed more than one client, the message might be in different stages of delivery for different clients. Therefore, more than one <cancel-result> can appear in each <cancel-response>. Each cancel-result can describe the result of more than one client. In this case, the <cancel-result> will contain an address element (refer to Figure 7.4)

```
<!ELEMENT cancel-result ( address* ) >
<!ATTLIST cancel-result
        code                    CDATA               #REQUIRED
        desc                    CDATA               #IMPLIED
>
```

Figure 7.11 The <cancel-result> element.

Table 7.2 Push Result Codes

STATE	MEANING
rejected	The message was not accepted.
pending	The message is in the process of being delivered.
delivered	The message was successfully delivered.
undeliverable	Some problem prevented the message from being delivered.
expired	The message could not be delivered in the specified time.
aborted	The client aborted the message.
timeout	The PPG's delivery process timed out.
cancelled	The message was cancelled by a cancel operation.

for each client that it represents. If no addresses are specified, the <cancel-result> pertains to all clients associated with the message. Figure 7.11 shows the DTD fragment that defines the <cancel-result> element.

The <cancel-result> element takes two attributes: `code` and `desc`. As with the `response-result` element, the `code` attribute indicates the status code of a cancel request (refer to Table 7.1), while the `desc` attribute contains a textual description of the code. Figure 7.11 shows the DTD fragment that defines the <cancel-result> element.

The <resultnotification-message> Element

The PPG uses the <resultnotification-message> element to report the final results of a message's delivery for a particular addressee to the PI. This element has eight attributes that are used to describe the results: `push-id`, `sender-address`, `sender-name`, `received-time`, `event-time`, `message-state`, `code`, and `desc`. The `push-id` attribute has the value of the original <push-message> ID. The `sender-address` and `sender-name` attribute has the values of the respective address and textual name of the source PPG. The `received-time` is a timestamp for when the PPG received the original `push-message`. The `event-time` is a timestamp for when the message achieved its final state. The `message-state` attribute specifies the final state of the attribute (see Table 7.2). The `code` and `desc` attributes have values for the status code (refer to Figure 7.1) and a textual description, respectively.

The <resultnotification-message> must contain an <address> element that defines the address of the client associated with the result. Furthermore, if a <quality-of-service> element was included in the original <push-message>, the PPG must include a quality-of-service element. This element describes the actual quality of service used to deliver the message. Figure 7.12 shows the DTD fragment that defines the <resultnotification-message> element.

The <resultnotifcation-response> Element

The <resultnotification-response> element is sent by the PI to the PPG in order to confirm receipt of a <resultnotification-message>. Figure 7.13 shows the DTD fragment that defines the <resultnotification-response> element.

The required <address> element and push-id attribute have the same value as in the <resultnotification-message>. These values are used by the PPG to match the response with the message. The code and desc attributes indicate the status code of the PI when receiving the message.

```
<!ELEMENT resultnotification-message ( address, quality-of-service? )
>
<!ATTLIST resultnotification-message
          push-id               CDATA          #REQUIRED
          sender-address        CDATA          #IMPLIED
          sender-name           CDATA          #IMPLIED
          received-time         %Datetime;     #IMPLIED
          event-time            %Datetime;     #IMPLIED
          message-state         %State;        #REQUIRED
          code                  CDATA          #REQUIRED
          desc                  CDATA          #IMPLIED
>
```

Figure 7.12 The <resultnotification-message> element.

```
<!ELEMENT resultnotification-response ( address ) >
<!ATTLIST resultnotification-response
          push-id               CDATA          #REQUIRED
          code                  CDATA          #REQUIRED
          desc                  CDATA          #IMPLIED
>
```

Figure 7.13 The <resultnotification-response> element.

```
<!ELEMENT statusquery-message ( address* ) >
<!ATTLIST statusquery-message
          push-id                     CDATA              #REQUIRED
>
```

Figure 7.14 The <statusquery-message> element.

```
<!ELEMENT statusquery-response ( statusquery-result+ ) >
<!ATTLIST statusquery-response
          push-id                     CDATA              #REQUIRED
>
```

Figure 7.15 The <statusquery-response> element.

The <statusquery-message> Element

The <statusquery-message> element is used to solicit the PPG for the status of a message with a given ID associated with one or more client addresses. Figure 7.14 shows the DTD fragment that defines the <statusquery-message> element. The push-id attribute indicates the original <push-message> upon which status is sought. The <statusquery-message> can contain one or more optional <address> elements. Each address indicates a client (associated with the push-id). In this case, status will be returned for the original <push-message> for the clients specified. If no <address> elements are specified, status will be returned for all clients associated with the <push-message>.

The <statusquery-response> Element

The <statusquery-response> is sent by the PPG in response to the PI sending a <statusquery-message>. The element has the push-id attribute that matches the push-id attribute of the request. The <statusquery-response> contains one or more required <statusquery-result> elements. Figure 7.15 shows the DTD fragment that defines the <statusquery-response> element.

The <statusquery-result> element is used within the content of a <statusquery-response> to report the status of a subset of addresses. Four attributes are used to describe the status: event-time, message-state, code, and desc. The <event-time> attribute indicates the time that the message went into its current state. The <message-state> attribute indicates the current state of the message (refer to Table 7.2). The code attribute indicates the current status of the message (refer to Table 7.1), and the desc attribute contains a textual description of the status.

Because each <stausquery-result> element reports the state of a subset of addresses, one or more optional address elements can be contained within the element. If no address elements are present, the <statusquery-result> represents all clients associated with the push-id. If a <quality-of-service> element was included in the original push-message, the PPG is required to send a <quality-of-service> element describing the quality of the final disposition of the message. Figure 7.16 shows the DTD fragment that defines the <statusquery-result> element.

The <badmessage-response> Element

The <badmessage-response> element is sent by either the PPG to the PI (or vice-versa) in response to a poorly formatted or otherwise bad message. This element contains a single attribute to contain the fragment of the message that is offensive: bad-message-fragment. Figure 7.17 shows the DTD fragment that defines the <badmessage-response> element.

Notes on Using Push Technology

When working with the various facets of Java development, JSDK Servlets, servlet engines, WML, and WML Script, you can set up a usable, self-contained development environment on a Personal Computer (PC). As you have seen by now, you can get a great deal of development done without having much more than that. Working with Push technology is different, however, for a few reasons. First, Push requires a real WAP client and a PPG. Second, Push deals with security issues that are beyond the

```
<!ELEMENT statusquery-result ( address*, quality-of-service? ) >
<!ATTLIST statusquery-result
          event-time               %Datetime;          #IMPLIED
          message-state            %State;             #REQUIRED
          code                     CDATA               #REQUIRED
          desc                     CDATA               #IMPLIED
>
```

Figure 7.16 The <statusquery-result> element.

```
<!ELEMENT badmessage-response EMPTY >
<!ATTLIST badmessage-response
          bad-message-fragment   CDATA                 #REQUIRED
>
```

Figure 7.17 The <badmessage-response> element.

scope of this text. Third, Push technology is relatively new technology (relative to WML and WML Script).

As of this writing, I am not aware of any vendor who offers Push in its WAP gateways or WAP clients. By the time this book is published, I am sure that there will be implementations available. Due to the security complications associated with Push technology, I believe that most developers will want to use the Software Development Kits (SDK) provided by PPG developers. Because they should all use the DTD described earlier, you should be able to select you favorite Java package when one is available.

Applications of Push Technology

While WAP technology augments the desktop experience by making the Internet more portable, Push technology bolsters WAP technology further by enabling WML decks to find you. In other words, rather than surfing to your favorite Web site for an update on your favorite content, you turn on the phone and the content comes to you.

In order to understand what a good application of Push technology is, you should first look at what does not make for good Push content. Unless users of WAP clients have paid lots of money for "all-you-can-eat" wireless service, they are paying by the minute for airtime. In some cases, users have to pay by the message for Push content. In either case, you can rest assured that users will not appreciate applications that generate unsolicited messages. Therefore, rule number 1 of Push technology is as follows: Do not push anything unless the user wants to have messages pushed. For example, special offers (such as upgrades for software products) and general "we haven't seen you at our site in a while" messages are best sent as e-mails, rather than as WAP Push messages.

Given that a contract of sorts should exist between the Web site and the end user, a good implementation of an application that pushes messages will enable users to select the type of message that they want pushed. For example, a news site should give users the capability to select the content that they want to be pushed out. Such a site should not push stock-market results to a user who is interested in seeing National Basketball Association (NBA) scores. This sort of granularity is good. Distinguishing between NBA and National Football League (NFL) scores, for example, would be even better. Picking the teams for which you want scores would be ideal. Applications that give a user a choice between surfing for data, receiving e-mail messages that include data of interest, or getting information pushed are viewed as flexible and useful.

If the user has requested a push of information, what type of Push should be used? As discussed earlier, if the contents of a message are time sensitive (for example, a meeting reminder), the message should include a `deliver-before-timestamp` attribute. If the user's WAP client is turned off during the time that a message is valid, the user will most likely not want to see the message after it is old.

Some messages contain volatile information. For example, an application that deals with stock quotes might enable a user to set up alarms on a particular stock when it is offered above or below a certain value. Such a message is not necessarily time sensitive, because the condition might exist for an indeterminate amount of time. The condition might change before the message is delivered to the WAP client, however. In such cases, the application must have the capability to cancel a message when the message becomes invalid due to a data change.

Continuing with the stock-quote example, the alarm event can be reasonably viewed as urgent relative to, say, a market-order confirmation. Therefore, this message might be sent with a high priority. If the example site offers the user the capability to purchase and/or sell holdings, the market-order confirmation should be sent at medium priority. This priority will enable the PPG to serve the alarm conditions more effectively than the order confirmations.

In most cases, if a message is important enough to send, the application should take care to make sure that it is delivered. To make sure that messages are delivered, the PI should register in order to receive confirmations. If for some reason (other than cancellation or expiration) a Push message is not delivered, a good site will be prepared to select an alternative message-delivery mechanism (perhaps e-mail). Alternatively, if the site permits users to receive periodic (perhaps hourly) updates on general information (market conditions, news, weather, traffic conditions, and so on), this information should be sent as low priority, with a `deliver-before-timestamp`, and delivered without confirmation.

JSDK Servlets and Push

Because this book deals with JSDK Servlets, we should mention how Push technology can be used in conjunction with a servlet. You can certainly set up an independent service as a PI. If your application needs the capability to service many users simultaneously and/or needs to be capable of handling many outstanding Push messages, I would recommend that you consider developing your PI as an independent Java application. You can implement your PI within your WAP Servlet if you so desire, however.

So far, we have discussed servlets as components of a servlet engine in which they run. Servlets can launch multiple threads, however, that can perform tasks independently of the servlet's normal behavior. While the *Central Processing Unit* (CPU) time used by these threads is borrowed from your servlet engine's process (a negative), a thread running in conjunction with a servlet can communicate with the servlet in memory with method calls (a plus). For smaller services with a minimal number of outstanding Push requests, such a thread can easily support a PI.

Conclusion

In this chapter, we discussed the WAP Push technology for delivery of asynchronous messages to the user WAP client. This technology, if used correctly, can greatly enhance the services of a site. In the next chapter, we will look at another WAP technology that has the capability to improve the performance of the WAP client: WML Script.

CHAPTER 8

Wireless Markup Language (WML) Script

In previous chapters, we saw how powerful the combination of WML and *Java Servlet Development Kit* (JSDK) servlets can be. Despite its flexibility, however, WML is still limited—because control is driven from a remote servlet. How much more powerful would a servlet be if it could actually send code to be executed on the Wireless Application Protocol (WAP) client device itself? In this chapter, we will examine WML Script and answer this question.

A Sample Application

Let's start with a simple variation on the *Hello World* theme. Like all Hello World applications, the following sample is straightforward—but it will serve as a good foundation and basic reference model.

Before you start developing WML Script servlets, you need to make sure that your server has a *Multi-Purpose Internet Mail Extensions* (MIME) type entry for the file extension `wmls`. If you review the `properties.txt` file located in the root directory of the `/code` directory on the accompanying CD-ROM in this book, you will note that I have already made an entry for both source and compiled WML Script code. Therefore, if you are using the Servlet engine that came with the book, you are all set.

Most of the time, you will reference a WML Script function from a WML document. Figure 8.1 shows a simple WML document. This WML document consists of two cards: card1 and card2. On card1, the *accept* soft key is associated with a hypertext reference to a WML Script function (the `SayHello()` function within the `sample.wmls` file). The soft key also initializes the variable `$(var1)` to the value *default*. You should notice that card2 is not directly referenced from card1; however, as we will see shortly,

```
<?xml version="1.0"?>
<!DOCTYPE wml PUBLIC "-//WAPFORUM//DTD WML 1.1//EN"
"http://www.wapforum.org/DTD/wml_1.1.xml">
<wml>
  <template>
    <do type="prev" label="Back">
      <prev/>
    </do>
  </template>
  <card title="WMLScript SAMPLE" id="card1">
    <onevent type="onenterforward">
      <refresh>
        <setvar name="var1" value="default"/>
      </refresh>
    </onevent>
    <do type="accept">
      <go href="sample.wmls#SayHello()"/>
    </do>
    <p align="center"><b>WMLScript Sample</b></p>
  </card>
  <card title="WMLScript Card2" id="card2">
    <p>$(var1)</p>
  </card>
</wml>
```

Figure 8.1 A sample WML document.

```
extern function SayHello()
{
  WMLBrowser.setVar("var1", "Hello, world!");
  WMLBrowser.go("fig02.wml#card2");
}
```

Figure 8.2 A sample WML Script compilation unit.

card2 is referenced from the function. The value of the variable `$(var1)` is displayed on card2.

Figure 8.2 shows a compilation unit. In WML Script terms, a compilation unit is a file that contains one or more WML Script functions. In this case, the file contains only one function: `SayHello()`. The `SayHello()` function is prefixed with the WML Script keyword `function`. The use of the keyword `extern` denotes that you can call the function from outside its compilation unit (for example, via a URL—as shown in the previous WML document). The function makes two relatively straightforward calls to the `WMLBrowser` library. First, the `setVar()` function is called to set the value of a

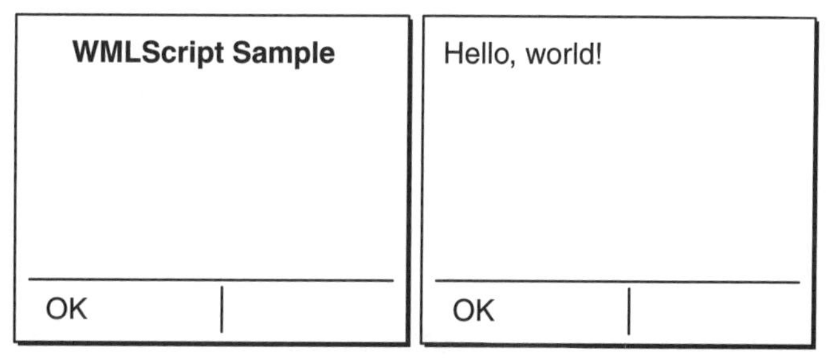

Figure 8.3 Output from the WML Script sample.

WML variable. Second, the `go()` function is called to force a new card to be loaded (card2 from our original WML deck).

You can find the sources of both of these files in the `/code/book/ chap08` directory on the accompanying CD-ROM. Figure 8.3 shows a progression of screen shots associated with the sample. Card1 appears in the first frame. By pressing the *accept* soft key on card1, you can display card2 via WML Script. You can access the WML deck for the sample via the following URL:

```
http://localhost/book/chap08/Fig02.wml
```

An Overview of WML Script

You could write an entire book about WML Script. Our purpose here, however, is to give you an understanding of its capabilities and a starting point for self-study. In this section, we will examine the language itself.

Lexical Structure

A lexical structure is a set of rules that establish how you write a program in a language. WML Script is a relatively new language. Additionally, many independent organizations are implementing their own version of the byte-code compilers and interpreters. Therefore, certain aspects of the lexical structure might be subject to interpretation. As with other areas of networking, developers should follow the robustness principle.

NOTE

In short, the robustness principle encourages individuals to be specific in their interpretation of a specification, but to be forgiving of others who might not be specific in their interpretation.

Case Sensitivity

In WML Script, reserved words and identifiers are all considered case sensitive. For robustness purposes, you should avoid using two or more identifiers that differ only in case. While it is legal and valid to do so, your script might someday find an implementation that does not handle case sensitivity correctly.

White Space and Line Breaks

As with other structured languages, you are free to break up lines and introduce white space whenever you feel that it will improve the readability of your code. White space within a literal is significant. Furthermore, a line break within a literal constitutes an error.

Optional Semicolons

Unlike Java Script, semicolons are required at the end of WML Script statements. Even if the WAP SDK or client that you are using happens to enable you to drop the semicolon, some implementations will produce an error.

Comments

You can use two forms of comments in WML Script. For short comments, you can use the single-line form that starts with a double forward slash (//) and terminates with a new line. When one or more comment lines are required, you can use the block-style comment that starts with a forward slash and asterisk (/*) and ends with an asterisk and forward slash (*/).

Literals

In WML Script, a literal is any value that appears directly in the code (for example, hard-coded values). There are literals for the following types: *integer*, *float*, *boolean*, and *strings*. You can express an integer literal in base-10 (no adornment), octal (with a 0 prefix), or hexadecimal (with a 0x) prefix. You can express floats as base-10 decimals (for example, 3.1416) or as exponentials (for example, 1.282E04). The Boolean literals are the keywords *true* and *false*. String literals are expressed as strings of characters enclosed in either double or single quotes. Table 8.1 shows the list of special escape characters that might appear within a string literal.

Identifiers

In WML Script, an identifier is any name that identifies a function, a variable, or a library. Identifiers can consist of one or more of the following characters: digits (0

Table 8.1 Special Escape Characters for String Literals

SEQUENCE	INTERPRETATION
\'	Single quote
\"	Double quote
\\	Backslash
\/	Forward slash
\b	Backspace
\f	Form feed
\n	New line
\r	Carriage return
\t	Horizontal tab
\xhh	ISO8859-1 character represented by the hexadecimal value hh
\ooo	ISO8859-1 character represented by the octal value ooo
\uhhhh	Unicode character represented by the hexadecimal value hhhh

Table 8.2 WML Script Reserved Words

access	agent	break	case*	
catch**	class**	const**	continue	
debugger**	default**	delete*	do**	
domain	else	enum**	equiv	export
extends**	extern	finally**	for	
function	header	http	if	
import**	in*	isvalid	lib*	
meta	name	new*	null*	
path	private**	public	return	
sizeof**	struct**	super**	switch**	
this*	throw**	try**	typeof	
url	use	var	void*	
while	with*			

*Keywords not used in WML Script.

**Keywords reserved for future use in WML Script.

through 9), letters (a through z and A through Z), and/or an underscore (_). An identifier cannot start with a number; otherwise, an identifier can be any name that does not conflict with a reserved word. Table 8.2 shows a list of WML Script reserved words.

TIP

While it is acceptable for literals with the same name to serve different purposes within the same context (for example, a variable and a function with the same name), I try to avoid this practice in my own work.

Variables

All WML Script variables must be declared within the context of a function prior to their first use. You declare a variable by preceding it with the keyword `var`. You can initialize variables at declaration time by using the assignment operator. The initial value can be either an expression or a literal. Figure 8.4 shows a code fragment in which variables are declared.

NOTE

As with other scripting languages, WML Script has variables that take on their type based on their use.

Functions

You must declare all functions within a *compilation unit*. To declare an identifier as a function, you must use the reserved word `function`. Normally, a function is limited to use within its compilation unit; however, the reserved word `extern` can precede the function declaration. When a function is declared external by using this method, you can access it either within or from the outside of its compilation unit.

As with other languages, WML Script functions can take arguments and return a value at the developer's discretion. If a function is to take arguments, a list of variables is provided within the parentheses of the function declaration. Alternatively, if a function does not take arguments, the argument list is left empty. All WML Script functions return a value. The type of the value returned is not declared. An implied value of an empty string is returned if a value is not explicitly returned by a function.

Figures 8.5a and 8.5b show a WML deck and a WML Script file, respectively. The WML deck invokes the WML Script. On the first card of the deck (card1), the user will select

```
function DemoFunction()
{
  var strValue;
  var strValue2 = "this is a test";
  var bValue = true;
  var nValue = 1;
  var nValue2 = nValue + 1;
  ...
}
```

Figure 8.4 Variable declaration.

```
<?xml version="1.0"?>
<!DOCTYPE wml PUBLIC "-//WAPFORUM//DTD WML 1.1//EN"
"http://www.wapforum.org/DTD/wml_1.1.xml">
<wml>
  <card id="card1">

    <onevent type="onenterforward">
      <refresh>
        <setvar name="var1" value="1"/>
        <setvar name="retval" value="Nothing picked"/>
      </refresh>
    </onevent>

    <do type="prev" label="Back">
      <prev/>
    </do>

    <do type="accept">
      <go href="fig5.wmls#Demo($var1) "/>
    </do>

    <p>Pick a number
      <select iname="var1">
        <option>one</option>
        <option>two</option>
        <option>three</option>
        <option>four</option>
      </select>
    </p>
  </card>
  <card id="card2">
    <do type="prev" label="Back">
      <prev/>
    </do>
    <p>$(retval)</p>
  </card>
</wml>
```

Figure 8.5a A function declaration example in `fig5.wml`.

an item from a WML select. When you press the *accept* soft key, the index of the selected item (stored in the variable `$(var1)`) is used as the only argument to an invocation of the `Demo()` function contained within the WML Script compilation module.

Within the `Demo()` function, if the index was 1, the function `DemoFunction1()` is invoked. Otherwise, the function `DemoFunction2()` is invoked with the index variable passed as an argument. `DemoFunction1()` returns a simple string to `Demo()` while `DemoFunction2()` returns a calculated string. The `Demo()` function sets the

```
extern function Demo(nPicker)
{
  var retval;

  if (nPicker == 1)
  {
    retval = DemoFunction1();
  }
  else
  {
    retval = DemoFunction2(nPicker);
  }
  WMLBrowser.setVar("retval", retval);
  WMLBrowser.go("fig5.wml#card2");
}

function DemoFunction1()
{
  return "You picked one";
}

function DemoFunction2(nValue)
{
  return "You picked " + nValue;
}
```

Figure 8.5b A function declaration example in `fig5.wmls`.

value of the `$(result)` WML variable to the value returned and returns control to card2 of the WML deck.

Libraries

Like many other languages (such as C, C++, and Java), much of the functionality of WML Script is found in standard libraries, rather than built into the language itself. There is a standard set of libraries built into each browser: Lang, Float, String, URL, WMLBrowser, and Dialogs. In the following sections, we will examine these libraries in more detail.

Lang Library

Do not be fooled by the name. This library looks more like Java's `math` package than its `lang` package. Table 8.3 lists the names of the functions in this library along with a brief description.

Table 8.3 `Lang` Library Functions<$Lang llibrary functions; WML Script>

FUNCTION	DESCRIPTION
`abs()`	Returns the absolute value of a number
`min()`	Returns the minimum of two values
`max()`	Returns the maximum of two values
`parseInt()`	Returns the integer representation of a string value
`parseFloat()`	Returns the float representation of a string value
`isInt()`	Returns true if the value of a string is an integer
`isFloat()`	Returns true if the value of a string is a float
`maxInt()`	Returns the maximum integer value supported by the browser implementation
`minInt()`	Returns the minimum integer value supported by the browser implementation
`float()`	Returns true if the browser implementation supports floats
`exit()`	Ends execution of the bytecode interpreter (normal exit condition)
`abort()`	Ends execution of the bytecode interpreter (abnormal exit condition)
`random()`	Returns a random number between zero and a specified value (inclusive)
`seed()`	Initializes the pseudo-random number generator

Table 8.4 Float Library Functions

FUNCTION	DESCRIPTION
`int()`	Returns the integer part of a float
`floor()`	Returns the integer value of a float rounding down
`ceil()`	Returns the integer value of a float rounding up
`pow()`	Returns the value of one number raised to the power of another
`sqrt()`	Returns the square root of a number
`maxFloat()`	Returns the maximum floating point value supported by the browser
`minFloat()`	Returns the minimum floating point value that the browser supports

Float Library

The Float library contains functions that are useful when dealing with floating point values. Table 8.4 lists the names of the functions in this library, along with a brief description.

Table 8.5 String Library Functions<$String Ilibrary functions; WML Script>

FUNCTION	DESCRIPTION
`length()`	Return the length of a string
`isEmpty()`	Returns true if a string is empty
`charAt()`	Returns the character at a specified position in a source string as a string
`subString()`	Returns a string by index and length from a source string
`find()`	Finds the first occurrence of a string within another string
`replace()`	Replaces all occurrences of an old value with a new value in a source string
`elements()`	Returns the number of delimited elements in a source string
`elementAt()`	Returns an element by index
`removeAt()`	Removes an element by index
`replaceAt()`	Replaces an element by index
`squeeze()`	Replaces all groups of consecutive white space with a single space
`trim()`	Returns a version of a source string with all leading and trailing white space removed
`compare()`	Compares one string to another
`toString()`	Converts a given non-string value to a string
`format()`	Creates a string with a C-style string format

String Library

The String library contains functions that are useful when dealing with string values. Table 8.5 lists the names of the functions in this library, along with a brief description.

NOTE

Like the Java `String` class, functions in the String library do not manipulate source strings; rather, a new String value (where appropriate) is returned.

URL Library

The URL library contains functions that are useful when dealing with URLs. Table 8.6 lists the names of the functions in this library, along with a brief description.

WMLBrowser Library

The WMLBrowser library contains functions that are useful when dealing with WML-Browsers. Table 8.7 lists the names of the functions in this library, along with a brief description.

Table 8.6 URL Library Functions

FUNCTION	DESCRIPTION
`isValid()`	Validates a given URL
`getScheme()`	Returns the scheme of the URL (for example, HTTP)
`getHost()`	Returns the host name portion of a URL
`getPath()`	Returns the path portion of a URL
`getParameters()`	Returns the parameters of a URL
`getQuery()`	Returns the query portion of a URL
`getFragment()`	Returns the anchor portion of a URL
`getBase()`	Returns the absolute URL
`getReferer()`	Returns the calling URL
`resolve()`	Resolves a URL from a base and an embedded URL
`escape()`	Returns the original URL with hex escape codes
`unescape()`	Returns a URL from an escaped URL
`escapeString()`	Fully escapes a string
`unescapeString()`	Fully unescapes a string
`loadString()`	Loads a string with the content associated with a URL

Table 8.7 WMLBrowser Library Functions

FUNCTION	DESCRIPTION
`getVar()`	Gets the value of a WML variable
`setVar()`	Sets the value of a WML browser
`go()`	Goes to (loads) a given hypertext reference
`prev()`	Executes a previous (as in a WML <prev>) on the browser
`newContext()`	Clears the browser context
`getCurrentCard()`	Returns the smallest relative URL to the card that is currently loaded in the browser
`refresh()`	Executes a refresh

NOTE

The `go()` and `prev()` functions are not immediate. In both cases, no action is taken until control is returned to the browser. Due to this delay, these functions override each other and themselves. In other words, you can call them multiple times within a script, but only the last call is executed upon return to the browser.

Table 8.8 Dialogs Library Functions

FUNCTION	DESCRIPTION
`prompt()`	Displays a message and the default value and returns an entered string
`confirm()`	Displays a message and returns true if accepted
`alert()`	Displays a message and waits for acknowledgement

Dialogs Library

The Dialogs library provides three convenience functions for displaying information in a dialog representation. Table 8.8 lists the names of the functions in this library, along with a brief description.

External Functions

From time to time, developers will want to have one compilation unit invoke a function in another compilation unit. In order to access such an external function, the calling compilation unit needs to effectively import the called compilation unit. You can perform this task with the keywords `use url`. The `use url` statement takes two parameters. The first parameter is an identifier that serves as the symbolic name of the external compilation unit. The second argument is a URL that identifies the source or binary WML Script file. Figure 8.6a shows a WML document consisting of one card. Pressing the *accept* soft key on this card invokes the external function `main()` within the compilation unit shown in Figure 8.6b. The `main()` function uses the compilation unit shown in Figure 8.6c, which contains the function `display()`. Figure 8.6d shows the screen shot that is displayed when the following URL is invoked:

```
http://localhost/book/chap08/fig6a.wml
```

Expressions

Expressions in WML Script are virtually the same as in Java. You can assign a variable the value of an expression. Furthermore, you can use an expression wherever a value could be specified. Table 8.9 shows a list of single type operators. Table 8.10 shows a list of multi-type operators.

Code Blocks

As with Java, code blocks are used to group related lines of code between a pair of braces ({ }). Code blocks are primarily used to associate lines with the various control statements if, else, while, and for.

NOTE

Unlike Java, variables defined within a code block cannot be defined again within another code block.

```
<?xml version="1.0"?>
<!DOCTYPE wml PUBLIC "-//WAPFORUM//DTD WML 1.1//EN"
"http://www.wapforum.org/DTD/wml_1.1.xml">
<wml>
  <card>

    <do type="accept">
      <go href="fig6b.wmls#main()"/>
    </do>

    <p>Press OK</p>
  </card>
</wml>
```

Figure 8.6a The WML document that invokes the example.

```
use url fig6c "http://localhost/book/chap08/fig6c.wmls";

extern function main()
{
  fig6c#display("Do you see how this works?");
}
```

Figure 8.6b The WML script that calls an external function.

```
extern function display(strMessage)
{
  Dialogs.confirm(strMessage,"yes","no");
}
```

Figure 8.6c The WML script that contains an external function.

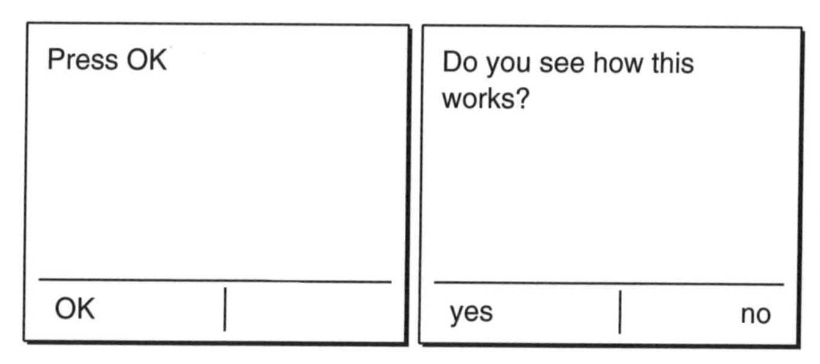

Figure 8.6d Screen shots from the external function example.

Table 8.9 Single Type Operators

OPERATOR	DESCRIPTION
!	Logical NOT
&&	Logical AND
||	Logical OR
~	Bitwise NOT
<<	Bitwise zero fill shift left
>>	Bitwise sign extended shift right
>>>	Bitwise zero fill shift right
&	Bitsize AND
^	Bitwise XOR
|	Bitwise OR
%	Integer remainder
div	Integer division
<<=	Bitwise zero fill shift left with assignment
>>=	Bitwise sign extended shift right with assignment
>>>=	Bitwise zero fill shift right with assignment
&=	Bitwise AND with assignment
^=	Bitwise XOR with assignment
|=	Bitwise OR with assignment
%=	Integer remainder with assignment
div=	Integer division with assignment

Control Statements

As with other languages, control statements can be used within WML Script to change the flow of control. In the following sections, we describe the control statements `if` and `else`.

If Statement

The `if` statement optionally executes a statement based on a Boolean value. As with Java, an `else` statement can be used in conjunction with an `if` statement to provide an alternative. A code block can be used in place of an individual statement as the outcome of either an `if` or an `else` statement. The `if-else` statement has the following syntax:

```
if (boolean-expression) statement; else statement;
```

Table 8.10 Multi-type Operators

OPERATOR	DESCRIPTION
++	Pre/post increment
-	Pre/post decrement
+	Unary plus
-	Unary minus
*	Multiplication
/	Division
-	Subtraction
+	Addition
<	Less than
<=	Less than or equal to
>	Greater than
>=	Greater than or equal to
==	Equal
!=	Not equal
*=	Multiply with assignment
/=	Division with assignment
-=	Subtraction with assignment
+=	Addition with assignment
typeof	Returns internal data type
isvalid	Check for validity
? :	Conditional expression
=	Assignment
,	Multiple evaluation

While Statement

The `while` statement optionally executes a statement as long as a Boolean expression is *true*. A code block can be used in place of the outcome statement. The `while` statement has the following syntax:

```
while (boolean-expression) statement;
```

For Statement

The `for` statement optionally executes a statement as long as a Boolean expression is *true*. As with Java, the `for` statement contains an optional initialize statement and an optional iteration statement. Likewise, the Boolean condition expression is optional. The `for` statement has the following syntax:

```
for (initialization; expression; iteration) statement;
```

Break Statement

A `break` statement is used to terminate the current (innermost) while or for loop and continue script execution with the statement following the loop's statement or code block.

Continue Statement

A `continue` statement is used to terminate the current (innermost) while or for loop and continue script execution with the next iteration of the loop. In the case of a `for` loop, the iteration statement of the for is executed and the condition is tested before the loop continues. In the case of a `while` loop, the condition is tested before the loop continues.

Return Statement

You can use a return statement at any point within the body of a function to return a value to the caller. Invocation of a return statement effectively terminates a function. If no return value is specified, a default value of an empty string is returned to the caller.

An Example: WML Script Blackjack

In previous sections, we looked at the basic elements of WML Script writing. In this section, we will reinforce what we have learned by developing a game as a WML Script. The game that we will develop is blackjack. What makes blackjack a good choice for an example in a professional text? First, it requires a good bit of string manipulation and math. Both of these areas happen to be strengths of WML Script and a weakness of WML alone.

The Requirements

The goal of this application is to demonstrate the capabilities of WML Script and how it can strengthen the WML that it accompanies. Note that this application is a simple implementation of the game. Therefore, if you are a blackjack aficionado, you might be disappointed. The following list describes the requirements for the application:

- The application should provide the notion of a bank. The bank gives the user the capability to track his or her winnings over several hands. Initially, the bank will contain $100.
- Prior to playing a hand, the user should be able to place a wager from his or her bank. The wager is specified in increments of $5, with a minimum bid of $5 and a maximum bid of $50.

- Once a wager is established, the player will see two cards from his or her hand and one card from the dealer's hand.
- The player will have the opportunity to receive another card (*hit*) or to stand with the cards shown. The user can choose to take additional cards until his or her hand exceeds 21 points (bust).
- Once a user stands or busts, the dealer will take additional cards until its hand has or exceeds 17 points.
- If the user wins, the value of the user's bid is added to the bank. If the user loses, the value of the bid is subtracted from the bank. If both players bust, no adjustment is made to the bank.

The Design

By using the Dialogs library, we can display simple messages. This capability, however, would not make for a pleasing display. Therefore, we need to have a fairly simple WML deck that we can use to display cards associated with bidding, game play, and results. The following design elements deal with the WML deck:

- We will use five variables in the WML deck: `$player`, `$dealer`, `$bank`, `$bid`, and `$result`. The `$player` variable will contain the cards in the human player's hand at any given time. The `$dealer` variable will contain the cards in the dealer's hand. The `$bank` variable will contain the number of dollars in the player's bank. The `$bid` variable will contain the dollar amount of the current bid. The `$result` variable will contain the final result of the hand.
- The first card will be a splash card displaying the name of the application and the author. Selecting the *accept* soft key will invoke the `deal()` function of the script (described as follows).
- A second card (ID new) will display the current `$bank` value as well as the current `$bid` value. There will be a link that will enable the user to play a hand based on the current bid. There will also be two links that will enable the user to *increment* and *decrement*, respectively, the current dollar amount of the `$bid` variable by $5. Selecting the *increment* link will invoke the `inc()` function of the script (described as follows). Selecting the *decrement* link will invoke the `dec()` function of the script (described as follows). You should note that upon initially entering this card, an *onenterforward* event will initialize the `$bank` variable to 100 and the `$bid` variable to 5.
- A third card (ID play) will display the current values for both the `$player` and `$dealer` variables. This card will also provide two links: *hit* and *stand*. Selecting the *hit* link will invoke the `hit()` function of the script (described as follows). Selecting the *stand* link will invoke the `stand()` function of the script. You should note that until the player stands or busts, this card will display the progress of the hand.
- A fourth card (ID result) will display the final state of a hand by displaying both the `$dealer` and `$player` variables along with the `$results` variable.

As noted earlier, all of the links established in the WML deck invoke functions in a WML Script. The following design elements pertain to the WML script:

- The `inc()` function will increment the `$bid` variable until it either exceeds the amount of money in the bank or until it reaches $50. If the value exceeds the amount in the bank, an alert dialog is displayed in order to notify the user. If the value exceeds $50, an alert dialog is displayed. Otherwise, the `$bid` variable is adjusted and the `new` card (the bidding card) is refreshed.
- The `dec()` function will decrease the `$bid` variable until it reaches $5. If the user attempts to call `dec()` while the current value of the `$bid` variable is $5, an alert dialog is displayed in order to notify the user. Otherwise, the `$bid` variable is adjusted and the `new` card is refreshed.
- The `deal()` function places a card at both the `$dealer` and `$player` variables and then invokes the `hit()` function. While in a real game of blackjack both players would be dealt two cards, only one card is dealt to each user in this function. Because the dealer only shows one card, the player only needs one card. The player is only dealt one card because the `hit()` function is called. (The `hit()` function will deal the second card to the player.)
- The `hit()` function places a card in the player's hand. At this point, one of two conditions will exist. Either the player's hand is bust, or it is not. If the player's hand is bust (more than 21 points), cards are dealt to the dealer's hand until the value of that hand is equal to or greater than 17. If the dealer does not bust, the win is awarded to the house and the player's bank is decremented by the amount of the current wager. If the dealer's hand is bust, the bank is not adjusted. With either dealer outcome, control is passed back to the WML deck at the result card. If the player's hand did not bust, the variables are updated and control is passed to the play card in the WML deck.
- The `stand()` function will calculate the results of the hand. At this point, we know that the player's hand is at or below 21 points. Cards are dealt to the dealer's hand until the value of the hand is equal to or greater than 17. If the dealer is bust or the value of the player's hand is greater than the dealer's, the hand is awarded to the player. If the dealer's hand is not bust and is greater than or equal to the player's hand, the hand is awarded to the house. In either case, the bank is adjusted appropriately—and control passes to the result card of the WML deck.
- The result card of the WML deck displays the final hands along with their point value (or the word BUST in the case of a bust). The results of the hand are also displayed ("House wins," "You win!," or "Push" in the case of a tie). The *accept* soft key of this card is associated with the new card; therefore, when the key is pressed, a new hand begins with the establishment of the next wager.

The Implementation

Based on the requirement and design elements stated in the previous sections, let's look at the implementation of the WML deck and WML servlet. The WML deck is shown in

```
<?xml version="1.0"?>
<!DOCTYPE wml PUBLIC "-//WAPFORUM//DTD WML 1.1//EN"
"http://www.wapforum.org/DTD/wml_1.1.xml">
<wml>
  <card>
    <onevent type="onenterforward">
      <refresh>
        <setvar name="bank" value="100"/>
        <setvar name="bet" value="5"/>
      </refresh>
    </onevent>
    <do type="accept">
      <go href="#new"/>
    </do>
    <p><b>Blackjack</b></p>
    <p>by John Cook</p>
  </card>
  <card id="new">
    <p>Bank:$$$(bank)
    <br/><a href="blackjack.wmls#deal()">Bet:$$$(bet)</a>
    <br/><a href="blackjack.wmls#inc()">+5</a>
    <br/><a href="blackjack.wmls#dec()">-5</a></p>
  </card>
  <card id="play">
    <p>Dealer:$(dealer)
    <br/>Player:$(player)
    <br/><a href="blackjack.wmls#hit()">Hit</a>
    <br/><a href="blackjack.wmls#stand()">Stand</a></p>
  </card>
  <card id="result">
    <do type="accept">
      <go href="#new"/>
    </do>
    <p>Dealer:$(dealer)
    <br/>Player:$(player)
    <br/>$(results)</p>
  </card>
</wml>
```

Figure 8.7 Source of the file `blackjack.wml`.

Figure 8.7. The source of this file resides at the following location on the accompanying CD-ROM:

```
/code/book/chap08/blackjack.wml
```

The WML deck consists of four cards: the default card, the new card, the play card, and the result card. The default card initializes two variables: `$(bank)` and `$(bet)`. The

`$(bank)` variable is initialized to 100, as in $100. The `$(bet)` variable is initialized to 5, as in $5. The *accept* soft key is associated with the new card. The display portion of the default card shows the title of the application and the author (splash screen).

The new card displays the current value of the bank. Three hypertext links are also shown. First, the value of the current bid is shown in a link to the `deal()` function of the WML script. Second, the text `+5` (as in, "Add $5 to my wager") is shown in a link to the `inc()` function of the WML script. Finally, the text `-5` (as in, "Subtract $5 from my wager") is shown in a link to the `dec()` function of the WML script.

The play card displays the current content of the dealer's hand. On the play card, we will always have one card representing the dealer's card that is shown. Next, the current content of the player's hand is shown. This value will initially be two cards representing the two cards of the player that were dealt face up. This value will change during the course of play, however, and will be shown when the player presses the *hit* link. Next, the value *Hit* is displayed in a link to the `hit()` function of the WML script. Finally, the value *Stand* is displayed in a link to the `stand()` function of the WML script.

The result card displays the current content of the dealer and player's hands. The results of the hand are also shown. The results of thc hand are provided in the variable `$(results)`, which is supplied by either the `hit()` or the `stand()` function. The result card associates the new card of the WML deck with the *accept* soft key.

Next, we will look at the source of the WML script by functions. The source of this file resides at the following location on the accompanying CD-ROM:

```
/code/book/chap08/blackjack.wmls
```

The calc( ) Function

Figure 8.8 shows the source of the `calc()` function. The purpose of this function is to calculate the value of a given hand. Because this calculation is performed on both the player and the dealer's hands at various points in the script, it is worth putting this calculation in its own function.

NOTE

Unless code needs to be in a function, it is best to place the code inline. Function calls (especially those that cross compilation modules) take more time than executing inline statements. Because the main goal of using WML Script is to save time, you should favor speed over the code's elegance.

Rather than reviewing the function line by line, I will explain its operation in prose. First, the function establishes variables for an integer to hold the running total (`nTotal`), a Boolean to track whether an ace has been seen (`bAce`), a string to contain a playing card (`strCard`), and the integer length of the hand (`nLength`). Next, the func-

```
function calc(strHand)
{
  var nTotal = 0;
  var bAce = false;
  var strCard;
  var vLength = String.length(strHand);
  while (-vLength >= 0)
  {
    strCard = String.charAt(strHand, vLength);

    if (Lang.isInt(strCard))
    {
      nTotal += Lang.parseInt(strCard);
      continue;
    }
    else if (strCard == "A")
    {
      bAce=true;
      nTotal++;
      continue;
    }
    nTotal += 10;
  }

  if ((nTotal <= 10) && bAce)
    nTotal += 10;

  return nTotal;
}
```

Figure 8.8 The `calc()` function of the WML script.

tion will loop through the hand, examining each playing card. If the value of the playing card is numeric (2 through 9, in the case of playing cards), its value is added to the running total. If the value of the card is an ace ("A"), a one is added to the running total and the `bAce` variable is set to *true*. Otherwise, the card must be a face card ("T" (10), "J," "Q," or "K"), and a 10 is added to the running total. At the end of the loop, if the hand contained an ace and the hand will support counting that ace as 11, rather than 1 (as was originally counted), a 10 is added to the running total. The value of the running total is returned.

The inc( ) Function

The `inc()` function is used to increment the value of the wager during the betting portion of the game. Figure 8.9 shows the source code for this function. Initially, values are

```
extern function inc()
{
  var strBank = WMLBrowser.getVar("bank");
  var nBank = Lang.parseInt(strBank);

  var strBet = WMLBrowser.getVar("bet");
  var nBet = Lang.parseInt(strBet);

  if ((nBank-nBet) <= 0)
  {
    Dialogs.alert("That bet would exceed your account.");
    return;
  }

  if ((nBet + 5) > 50)
  {
    Dialogs.alert("Maximum bet is $50");
    return;
  }

  nBet += 5;
  WMLBrowser.setVar("bet", String.toString(nBet));

  WMLBrowser.refresh();
}
```

Figure 8.9 The `inc()` function of the WML script.

established for both the integer amount in the bank (`nBank`) and the current integer value of the wager (`nBet`).

NOTE

When the WML variables `$(bank)` and `$(bet)` are retrieved, they are returned as strings. Because we need to perform arithmetic operations on these values, they must be converted to integers via the `Lang.parseInt()` library function. Eventually, these values will be returned to their WML variables as strings, so another conversion will be required via the `String.toString()` library function.

A check is performed to determine whether the value of the new wager would exceed the amount in the bank. If so, an error is displayed in a dialog via the `Dialogs.alert()` library function, and control is returned to the new card of the WML deck. Another check is performed to determine whether the value of the new wager would exceed $50. If so, an error is displayed, and control returns to the new card. If both of the tests fail, the value of the wager is changed, the wager is stored back into its WML variable, and control is returned to the new card.

The dec( ) Function

The `dec()` function is used to decrement the value of the wager during the betting portion of the game. Figure 8.10 shows the source code for this function. Initially, values are established for both the integer amount in the bank (`nBank`) and the current integer value of the wager (`nBet`).

A check is performed to determine whether the value of the new wager would be zero. If so, an error is displayed in a dialog via the `Dialogs. alert()` library function, and control is returned to the new card of the WML deck. If the test fails, the value of the wager is changed, the wager is stored back into its WML variable, and control is returned to the new card.

The deal( ) Function

The wagering portion of the hand is concluded with a call to the `deal()` function. The source of this function is shown in Figure 8.11. In this function, both the dealer and the player are dealt one card. The dealer's hand is represented by the WML variable `$(dealer)`, while the player's hand is represented by the `$(player)` variable. Finally, the `hit()` function is called. Calling the `hit()` function will deal the second

```
extern function dec()
{
  var strBank = WMLBrowser.getVar("bank");
  var nBank = Lang.parseInt(strBank);

  var strBet = WMLBrowser.getVar("bet");
  var nBet = Lang.parseInt(strBet);

  if ((nBet-5) < 5)
  {
    Dialogs.alert("Minimum bet $5.");
    return;
  }

  nBet -= 5;
  WMLBrowser.setVar("bet", String.toString(nBet));

  WMLBrowser.refresh();
}
```

Figure 8.10 The `dec()` function of the WML script.

```
extern function deal()
{
  WMLBrowser.setVar("dealer",
    String.charAt("A23456789TJQK", Lang.random(12)));
  WMLBrowser.setVar("player",
    String.charAt("A23456789TJQK", Lang.random(12)));
  hit();
}
```

Figure 8.11 The `deal()` function of the WML script.

card to the player's hand, refresh the WML variables, and return control to the play card of the WML deck.

The hit( ) Function

During the play of the hand, the `hit()` function is called in response to the user selecting the *hit* link. The source of this function is shown in Figure 8.12. First, a card is dealt to the player's hand, and the value of the hand is calculated. If the value of the player's hand is less than or equal to 21, the work of the function is complete—and control is returned to the caller (the play card of the WML deck).

If the value of the player's hand is greater than 21, the hand is complete—and additional steps are taken to calculate the results. First, the word "BUST" is added to the player's hand. Next, the dealer's hand is finished. Finishing the dealer's hand consists of dealing enough cards to the dealer in order to establish a hand with a point value greater than or equal to 17. Once the dealer's hand is complete, the point value of the dealer's hand is calculated. At this point, the dealer has won if its point value is less than or equal to 21. In this case, the wager is subtracted from the bank, the value of the dealer's hand is appended to its hand, and the `$(results)` variable is set to "House wins." Otherwise, the dealer is bust as well. In this case, the bank is not adjusted, the word "BUST" is appended to the dealer's hand, and the `$(results)` variable is set to "Push."

The stand( ) Function

The `stand()` function is called when the player selects the stand link. The source of this function is shown in Figure 8.13. A stand effectively ends the hand. Because the last hit operation (if there was one) did not result in a bust, we know at this point that the player has a good hand; therefore, the point value of the player's hand is calculated and appended to the hand. Next, the dealer's hand is finished (as described earlier in the `hit()` function). At this point, if the dealer's hand is bust or the value of the dealer's hand is less than the value of the player's hand, the player wins. In this case, the bank is adjusted in favor of the player, and the `$(results)` variable is set to "You win!" Otherwise, the dealer wins. In this case, the bank is adjusted in the house's favor, and the `$(results)` variable is set to "House wins." In either case, the appropriate text (either the value of the hand or the word "BUST" is appended to the dealer's hand.

```
extern function hit()
{
  var strPlayer = WMLBrowser.getVar("player");

  strPlayer += String.charAt("A23456789TJQK", Lang.random(12));
  WMLBrowser.setVar("player", strPlayer);

  var strGo = "play";

  if (calc(strPlayer) > 21)
  {
    var strDealer = WMLBrowser.getVar("dealer");

    strPlayer += " BUST";
    WMLBrowser.setVar("player", strPlayer);

    var nDealer;

    for (;;)
    {
      nDealer = calc(strDealer);
      if (nDealer >= 17)
        break;
      strDealer += String.charAt("A23456789TJQK", Lang.random(12));
    }

    var strDealerResult = " BUST";
    WMLBrowser.setVar("results","Push");

    if (nDealer <= 21)
    {
      strDealerResult = String.toString(nDealer);
      WMLBrowser.setVar("results","House wins");
      var nBank = Lang.parseInt(WMLBrowser.getVar("bank"));
      var nBet = Lang.parseInt(WMLBrowser.getVar("bet"));
      nBank -= nBet;
      WMLBrowser.setVar("bank", String.toString(nBank));
      if (nBank < nBet)
      {
        nBet = nBank;
        WMLBrowser.setVar("bet", String.toString(nBet));
      }
    }

    strDealer += " " + strDealerResult;
    strGo = "result";
```

(continues)

Figure 8.12 The `hit()` function of the WML script.

```
    WMLBrowser.setVar("dealer", strDealer);
  }

  WMLBrowser.refresh();
  WMLBrowser.go("blackjack.wml#" + strGo);
```

Figure 8.12 The `hit()` function of the WML script (Continued).

Playing the Game

To execute the game, you should start the servlet engine that came with the accompanying CD-ROM. The WML deck for the blackjack application is located at the following URL:

```
http://localhost/book/chap08/blackjack.wml
```

Using your preferred WAP SDK simulator, load the URL. You will see the splash screen shown in the first panel of Figure 8.14. Once the *accept* soft key is pressed on the splash screen, the wager screen (corresponding to the WML deck card with an ID of new) appears. At this point, the user has the ability to increase, decrease, and play with the initial wager of $5. In the game shown in Figure 8.14, I chose to increase my wager to $10 (panel two) and selected “Bet: $10” to get the initial cards shown in panel three. With the dealer showing a jack and my hand containing six points, I elected to take a hit to get the results shown in panel four. With the addition of a 10, I decided to take another hit (a poor choice). As luck would have it, however, the dealer busted as well. The results of the hand are shown in panel five. Selecting the *accept* soft key on the result card placed me back on the wager screen, as shown in panel six.

Ramifications

By now, you have noticed that the blackjack example does not use a servlet. The focus of this chapter has been on writing WML script; however, because the book is about WAP servlets, we should at least discuss how you can use servlets in conjunction with WML Script. While the blackjack example exists in static files, it could certainly have been dynamically rendered in whole or in part by a servlet.

There are several good reasons why you might want to render portions of your WML script via a servlet. Consider the following list:

- Internationalization—Because a WML script can contain text and/or phrases, it might be nice to render language-appropriate text into the script as it is served to the client device. If a script deals with time, date, and/or currency (as with the blackjack example), this information can be taken into account as well.
- Licensing/fees—If your script application contains licensing restrictions and/or use fees, you can better control access and accounting by serving your scripts from a servlet that can validate and account for usage.

```
extern function stand()
{
  var strPlayer = WMLBrowser.getVar("player");
  var strDealer = WMLBrowser.getVar("dealer");

  var nPlayer = calc(strPlayer);
  strPlayer += " " + String.toString(nPlayer);

  var nDealer;

  for (;;)
  {
    nDealer = calc(strDealer);
    if (nDealer >= 17)
      break;
    strDealer += String.charAt("A23456789TJQK", Lang.random(12));
  }

  var nBank = Lang.parseInt(WMLBrowser.getVar("bank"));
  var nBet = Lang.parseInt(WMLBrowser.getVar("bet"));

  var strDealerResults = String.toString(nDealer);
  var strResults = "House wins";
  var nAdjust = -1;
  if ((nDealer > 21) || (nPlayer > nDealer))
  {
    strDealerResults = "BUST";
    strResults = "You win!";
    nAdjust = 1;
  }

  strDealer += " " + strDealerResults;
  nBank += nBet * nAdjust;

  WMLBrowser.setVar("results", strResults);
  WMLBrowser.setVar("bank", String.toString(nBank));

  if (nBank < nBet)
  {
    nBet = nBank;
```

(continues)

Figure 8.13 The `stand()` function of the WML script.

- Persistent script variables—To a certain extent, script variables are persistent within the client device; however, via the use of a servlet, variables (such as the bank account in the blackjack application) could be stored in a user profile

```
    WMLBrowser.setVar("bet", String.toString(nBet));
  }

  WMLBrowser.setVar("player", strPlayer);
  WMLBrowser.setVar("dealer", strDealer);
  WMLBrowser.refresh();
  WMLBrowser.go("blackjack.wml#result");
}
```

Figure 8.13 The `stand()` function of the WML script (Continued).

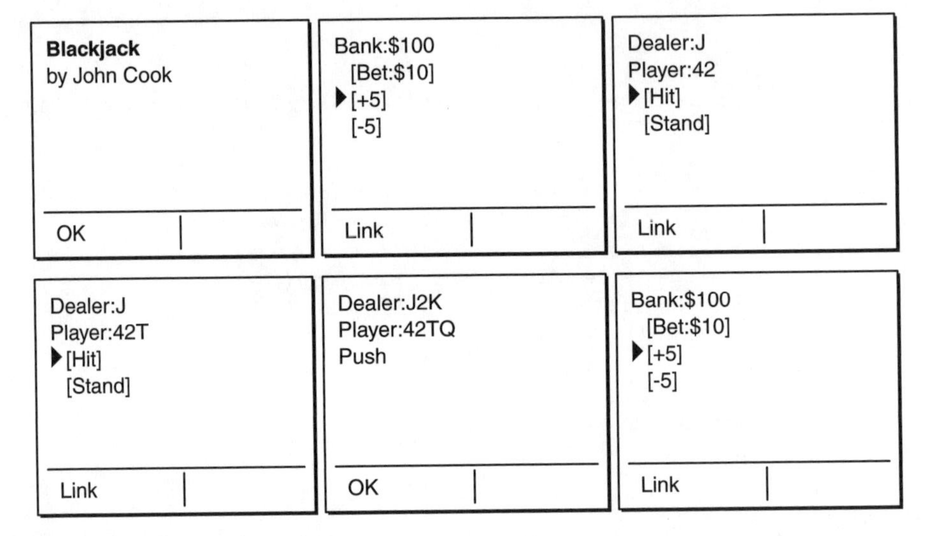

Figure 8.14 Game screen shots.

maintained by a servlet and associated with a cookie stored in the device. You should notice that using WML Script variables for persistent storage is not a good idea because they are erased when the browser is reset.

- Tailoring scripts—The real strength of WML Script is in its capability to validate WML deck data on the device before it is returned to a servlet for processing. Depending on the sophistication of your servlet, you might choose to tailor your validation scripts according to the data being entered. For example, if an application requires collecting credit card numbers, you might wish to change the content of your validation script to correspond to the type of card selected by the user.

Conclusion

In this chapter, we learned about how we can use WML Script to offload some routine calculations to the client device. With the growth of the Internet and the increasing demand placed on servers, remote calculations achieved with scripts can greatly reduce both the chattiness and computational requirements of a Web application.

As with WML, WML Script is an evolving technology. With many vendors developing WML Script byte-code interpreters for their devices, not all WML scripts will run the same on all devices. A final word of caution: Make sure that your applications work on the client devices that you intend to target.

APPENDIX A

Wireless Markup Language (WML) Reference

This appendix contains WML elements in reference form from the Wireless Application Protocol (WAP) 1.2 specification. Because the book's examples are based on WML 1.1, one element (<pre>) and a couple of attributes (accesskey and enctype) are new. For the most part, however, there were not a great deal of changes between WML 1.1 and WML 1.2.

In an attempt to simplify the reading of the Document Template Definition (DTD), I have removed many of the original eXtensible Markup Language (XML) entities and included their actual values inline. Therefore, you should be able to read individual sections of this appendix without a great deal of looking around to discover what an element actually does. Some macros from the original text are useful, however, in that they describe a particular convention that is implied (rather than called out explicitly) in the DTD. The following table defines those entities:

ENTITY	VALUE	MEANING
%length;	CDATA	Length expressed as a numeric value [0-9]+ or as a percentage [0-9]+"%"
%vdata;	CDATA	A string value that can contain variable references
%HREF;	CDATA	A URI, URL, or Uniform Resource Name (URN) designation that can contain variable references
%boolean;	(true|false)	A Boolean value

I also included a section for each element that describes where an element can be used. Therefore, if you see an element that you think you might like to use, you can quickly verify in which elements it is valid.

<a>

The <a> element is used to establish a link to a *Uniform Resource Locator* (URL). The text content of the element is displayed on the browser as a link.

DTD Definition

```
<!ELEMENT a ( #PCDATA | br | img )*>
<!ATTLIST a
    href          %HREF;          #REQUIRED
    title         %vdata;         #IMPLIED
    accesskey     %vdata;         #IMPLIED
    xml:lang      NMTOKEN         #IMPLIED
    id            ID              #IMPLIED
    class         CDATA           #IMPLIED>
```

Elements That Might Contain This Element

ELEMENT	USED FOR
<b>	Bold text
<big>	Larger-than-normal text
<em>	Emphasized text
<fieldset>	Group text and/or input fields
<i>	Italic text
<p>	Paragraph of text
<pre>	Preformatted text
<small>	Smaller-than-normal text
<strong>	Stronger-than-normal text
<td>	Data element within a table
<u>	Underlined text

<access>

The <access> element is used to restrict access to a WML deck to particular URLs.

DTD Definition

```
<!ELEMENT access EMPTY>
<!ATTLIST access
```

```
domain          CDATA           #IMPLIED
path            CDATA           #IMPLIED
id              ID              #IMPLIED
class           CDATA           #IMPLIED>
```

Elements That Might Contain This Element

ELEMENT	USED FOR
<head>	Establishes the behavior of a WML deck

<anchor>

The <anchor> element is used to establish a link to a URL. The text content of the element is displayed on the browser as a link. This element is a more flexible form of the <a> element.

DTD Definition

```
<!ELEMENT anchor (#PCDATA | br | img | go | prev | refresh )*>
<!ATTLIST anchor
    title           %vdata;          #IMPLIED
    accesskey       %vdata;          #IMPLIED
    xml:lang        NMTOKEN          #IMPLIED
    id              ID               #IMPLIED
    class           CDATA            #IMPLIED>
```

Elements That Might Contain This Element

ELEMENT	USED FOR
<b>	Bold text
<big>	Larger-than-normal text
<em>	Emphasized text
<fieldset>	Group text and/or input fields
<i>	Italic text
<p>	Paragraph of text
<small>	Smaller-than-normal text
<strong>	Stronger-than-normal text
<td>	Data element within a table
<u>	Underlined text

<b>

The <b> element is used to display text in a bold font.

DTD Definition

```
<!ELEMENT b (PCDATA | em | strong | b | i |
    u | b | small | br | img | anchor | a | table)*>
<!ATTLIST big
    xml:lang        NMTOKEN              #IMPLIED
    id              ID                   #IMPLIED
    class           CDATA                #IMPLIED>
```

Elements That Might Contain This Element

ELEMENT	USED FOR
<b>	Bold text
<big>	Larger-than-normal text
<em>	Emphasized text
<fieldset>	Group text and/or input fields
<i>	Italic text
<p>	Paragraph of text
<pre>	Preformatted text
<small>	Smaller-than-normal text
<strong>	Stronger-than-normal text
<td>	Data element within a table
<u>	Underlined text

<big>

The <big> element is used to display text larger than normal.

DTD Definition

```
<!ELEMENT big (PCDATA | em | strong | b | i |
    u | big | small | br | img | anchor | a | table)*>
<!ATTLIST big
    xml:lang        NMTOKEN              #IMPLIED
    id              ID                   #IMPLIED
    class           CDATA                #IMPLIED>
```

Elements That Might Contain This Element

ELEMENT	USED FOR
<b>	Bold text
<big>	Larger-than-normal text
<em>	Emphasized text
<fieldset>	Group text and/or input fields
<i>	Italic text
<p>	Paragraph of text
<small>	Smaller-than-normal text
<strong>	Stronger-than-normal text
<td>	Data element within a table
<u>	Underlined text

The <br/> element is used to insert a line break within text.

DTD Definition

```
<!ELEMENT br EMPTY>
<!ATTLIST br
    id          ID          #IMPLIED
    class       CDATA       #IMPLIED>
```

Elements That Might Contain This Element

ELEMENT	USED FOR
<a>	Link to a URL.
<anchor>	Alternate (more flexible form of <a>)
<b>	Bold text
<big>	Larger-than-normal text
<em>	Emphasized text
<fieldset>	Group text and/or input fields
<i>	Italic text
<p>	Paragraph of text
<pre>	Preformatted text

ELEMENT	USED FOR
<small>	Smaller-than-normal text
<strong>	Stronger-than-normal text
<td>	Data element within a table
<u>	Underlined text

<card>

The <card> element represents a virtual page within a WML document.

DTD Definition

```
<!ELEMENT card (onevent*, timer?, (do | p | pre)*)>
<!ATTLIST card
    title           %vdata;         #IMPLIED
    newcontext      %boolean;       "false"
    ordered         %boolean;       "true"
    xml:lang        NMTOKEN         #IMPLIED
    onenterforward  %HREF           #IMPLIED
    onenterbackward %HREF           #IMPLIED
    ontimer         %HREF           #IMPLIED
    id              ID              #IMPLIED
    class           CDATA           #IMPLIED>
```

Elements That Might Contain This Element

ELEMENT	USED FOR
<wml>	Root element of a WML deck

<do>

The <do> element is used to associate functions (usually represented by soft keys, hard keys, and/or browser menu items) with an action.

DTD Definition

```
<!ELEMENT do (go | prev | noop | refresh)>
<!ATTLIST do
    type            CDATA           #REQUIRED
    label           %vdata;         #IMPLIED
    name            NMTOKEN         #IMPLIED
```

```
    optional        %boolean;        "false"
    xml:lang        NMTOKEN          #IMPLIED
    id              ID               #IMPLIED
    class           CDATA            #IMPLIED>
```

Elements That Might Contain This Element

ELEMENT	USED FOR
<card>	Defines a virtual page within a WML deck
<fieldset>	Group text and/or input fields
<p>	Paragraph of text
<template>	Defines navigation elements that apply to all cards

<em>

The <em> element is used to display text with more emphasis than normal.

DTD Definition

```
<!ELEMENT em (PCDATA | em | strong | b | i |
    u | big | small | br | img | anchor | a | table)*>
<!ATTLIST em
    xml:lang        NMTOKEN          #IMPLIED
    id              ID               #IMPLIED
    class           CDATA            #IMPLIED>
```

Elements That Might Contain This Element

ELEMENT	USED FOR
<b>	Bold text
<big>	Larger-than-normal text
<em>	Emphasized text
<fieldset>	Group text and/or input fields
<i>	Italic text
<p>	Paragraph of text
<pre>	Preformatted text
<small>	Smaller-than-normal text
<strong>	Stronger-than-normal text

ELEMENT	USED FOR
<td>	Data element within a table
<u>	Underlined text

<fieldset>

The <fieldset> element is used to group text and/or input items within a WML card.

DTD Definition

```
<!ELEMENT fieldset (PCDATA | em | strong | b | i |
    u | big | small | br | img | anchor | a |
    table | input | select | fieldset | do)*>
<!ATTLIST fieldset
    title           CDATA                #IMPLIED
    xml:lang        NMTOKEN              #IMPLIED
    id              ID                   #IMPLIED
    class           CDATA                #IMPLIED>
```

Elements That Might Contain This Element

ELEMENT	USED FOR
<fieldset>	Group text and/or input fields
<p>	Paragraph of text

<go>

The <go> element is used to invoke a URL.

DTD Definition

```
<!ELEMENT go (postfield | setvar)*>
<!ATTLIST go
    href            %HREF;       #REQUIRED
    sendreferer     %boolean;    "false"
    method          (post|get)   "get"
    enctype         CDATA        "application/x-www-form-urlencoded"
    accept-charset  CDATA        #IMPLIED
    id              ID           #IMPLIED
    class           CDATA        #IMPLIED>
```

<head>

The <head> element contains <access> and <meta> elements that define the behavior of a WML deck.

DTD Definition

```
<!ELEMENT head ( access | meta )+>
<!ATTLIST head
    id              ID              #IMPLIED
    class           CDATA           #IMPLIED>
```

Elements That Might Contain This Element

ELEMENT	USED FOR
<wml>	Root element of a WML deck

<i>

The <i> element is used to display text in italics.

DTD Definition

```
<!ELEMENT i (PCDATA | em | strong | b | i |
    u | big | small | br | img | anchor | a | table)*>
<!ATTLIST i
    xml:lang NMTOKEN #IMPLIED
    id              ID              #IMPLIED
    class           CDATA           #IMPLIED>
```

Elements That Might Contain This Element

ELEMENT	USED FOR
<b>	Bold text
<big>	Larger-than-normal text
<em>	Emphasized text
<fieldset>	Group text and/or input fields
<i>	Italic text
<p>	Paragraph of text

ELEMENT	USED FOR
<pre>	Preformatted text
<small>	Smaller-than-normal text
<strong>	Stronger-than-normal text
<td>	Data element within a table
<u>	Underlined text

<img>

The <img> element is used to imbed an image within a WML document.

DTD Definition

```
<!ELEMENT img EMPTY>
<!ATTLIST img
    alt          %vdata;                #REQUIRED
    src          %HREF;                 #REQUIRED
    localsrc     %vdata;                #IMPLIED
    vspace       %length;               "0"
    hspace       %length;               "0"
    align        (top|middle|bottom)    "bottom"
    height       %length;               #IMPLIED
    width        %length;               #IMPLIED
    xml:lang     NMTOKEN                #IMPLIED
    id           ID                     #IMPLIED
    class        CDATA                  #IMPLIED>
```

Elements That Might Contain This Element

ELEMENT	USED FOR
<a>	Link to a URL
<anchor>	Alternate (more flexible form of <a>)
<b>	Bold text
<big>	Larger-than-normal text
<em>	Emphasized text
<fieldset>	Group text and/or input fields
<i>	Italic text
<p>	Paragraph of text
<small>	Smaller-than-normal text
<strong>	Stronger-than-normal text

<td>	Data element within a table
<u>	Underlined text

<input>

The <input> element is used to solicit input from an end user.

DTD Definition

```
<!ELEMENT input EMPTY>
<!ATTLIST input
    name          NMTOKEN            #REQUIRED
    type          (text|password)    "text"
    value         %vdata;            #IMPLIED
    format        CDATA              #IMPLIED
    emptyok       %boolean;          "false"
    size          %number;           #IMPLIED
    maxlength     %number;           #IMPLIED
    tabindex      %number;           #IMPLIED
    title         %vdata;            #IMPLIED
    accesskey     %vdata;            #IMPLIED
    xml:lang      NMTOKEN            #IMPLIED
    id            ID                 #IMPLIED
    class         CDATA              #IMPLIED>
```

Elements That Might Contain This Element

ELEMENT	USED FOR
<fieldset>	Group text and/or input fields
<p>	Paragraph of text
<pre>	Preformatted text

<meta>

The <meta> element is used to send meta commands to a WAP browser.

DTD Definition

```
<!ELEMENT meta EMPTY>
<!ATTLIST meta
    http-equiv    CDATA              #IMPLIED
    name          CDATA              #IMPLIED
    forua         %boolean;          #IMPLIED
```

```
    content         CDATA               #REQUIRED
    scheme          CDATA               #IMPLIED
    id              ID                  #IMPLIED
    class           CDATA               #IMPLIED>
```

Elements That Might Contain This Element

ELEMENT	USED FOR
<head>	Establishes the behavior of a WML deck

<noop>

The <noop> element performs no function. You can use this element to override the default or more general functions previously established in a <do> or <onevent> element.

DTD Definition

```
<!ELEMENT noop EMPTY>
<!ATTLIST noop
    id              ID                  #IMPLIED
    class           CDATA               #IMPLIED>
```

Elements That Might Contain This Element

ELEMENT	USED FOR
<do>	Associates a function with an action
<onevent>	Establishes the outcome of an event
<template>	Defines navigation elements that apply to all cards

<onevent>

The <onevent> element is used to establish the outcome of an event.

DTD Definition

```
<!ELEMENT onevent (go | prev | noop | refresh)>
<!ATTLIST onevent
    type            CDATA               #REQUIRED
    id              ID                  #IMPLIED
    class           CDATA               #IMPLIED>
```

Elements That Might Contain This Element

ELEMENT	USED FOR
<card>	Defines a virtual page within a WML deck
<option>	Defines an entry within a <select> element
<select>	Used to offer the end user multiple choices

<optgroup>

Using the <optgroup> element, you can group related <option> elements into a tree structure.

DTD Definition

```
<!ELEMENT optgroup (optgroup|option)+>
<!ATTLIST optgroup
    title           %vdata;            #IMPLIED
    xml:lang        NMTOKEN            #IMPLIED
    id              ID                 #IMPLIED
    class           CDATA              #IMPLIED>
```

Elements That Might Contain This Element

ELEMENT	USED FOR
<optgroup>	Groups related <option> elements
<select>	Used to offer the end user multiple choices

<option>

The <option> element is used to define an entry within a <select> element.

DTD Definition

```
<!ELEMENT option (#PCDATA | onevent)*>
<!ATTLIST option
    value           %vdata;            #IMPLIED
    title           %vdata;            #IMPLIED
    onpick          %HREF;             #IMPLIED
    xml:lang        NMTOKEN            #IMPLIED
    id              ID                 #IMPLIED
    class           CDATA              #IMPLIED>
```

Elements That Might Contain This Element

ELEMENT	USED FOR
<optgroup>	Groups related <option> elements

<p>

The <p> element is used to define a paragraph of text to be displayed on a WAP browser.

DTD Definition

```
<!ELEMENT p (#PCDATA | em | strong | b | i | u |
    big | small | br | img | anchor | a | table |
    input | select | fieldset | do)*>
<!ATTLIST p
    align          (left|right|center)  "left"
    mode           (wrap|nowrap)        #IMPLIED
    xml:lang       NMTOKEN              #IMPLIED
    id             ID                   #IMPLIED
    class          CDATA                #IMPLIED>
```

Elements That Might Contain This Element

ELEMENT	USED FOR
<card>	Defines a virtual page within a WML deck

<postfield>

The <postfield> element is used to define a name-value pair to be placed in the content section of an HTTP post request.

DTD Definition

```
<!ELEMENT postfield EMPTY>
<!ATTLIST postfield
    name           %vdata;              #REQUIRED
    value          %vdata;              #REQUIRED
```

```
    id              ID                  #IMPLIED
    class           CDATA               #IMPLIED>
```

Elements That Might Contain This Element

ELEMENT	USED FOR
<go>	Invokes a URL

<pre>

The <pre> element instructs the WAP browser to display the enclosed text as preformatted text. White space should be left intact, and line formatting should be honored.

DTD Definition

```
<!ELEMENT pre (#PCDATA | a | br | i | b| em | strong | input | select)*>
<!ATTLIST pre
    xml:space       CDATA               #FIXED "preserve"
    id              ID                  #IMPLIED
    class           CDATA               #IMPLIED>
```

Elements That Might Contain This Element

ELEMENT	USED FOR
<card>	Defining a card within a WML deck

<prev>

The <prev> element is used to force backward navigation of a WAP browser by popping the previous card off the browser's history stack.

DTD Definition

```
<!ELEMENT prev (setvar)*>
<!ATTLIST prev
    id              ID                  #IMPLIED
    class           CDATA               #IMPLIED>
```

Elements That Might Contain This Element

ELEMENT	USED FOR
<anchor>	Alternate (more flexible form of <a>)
<do>	Associates a function with an action
<onevent>	Establishes the outcome of an event

<refresh>

The <refresh> element is used to refresh the display of a WAP browser when the value of displayed variables has changed.

DTD Definition

```
<!ELEMENT refresh (setvar)*>
<!ATTLIST refresh
    id              ID                  #IMPLIED
    class           CDATA               #IMPLIED>
```

Elements That Might Contain This Element

ELEMENT	USED FOR
<anchor>	Alternate (more flexible form of <a>)
<do>	Associates a function with an action
<onevent>	Establishes the outcome of an event

<select>

The <select> element is used to offer the end user multiple choices from which one item can be selected.

DTD Definition

```
<!ELEMENT select (optgroup|option)+>
<!ATTLIST select
    title           %vdata;             #IMPLIED
    name            NMTOKEN             #IMPLIED
    value           %vdata;             #IMPLIED
    iname           NMTOKEN             #IMPLIED
    ivalue          %vdata;             #IMPLIED
    multiple        %boolean;           "false"
    tabindex        %number;            #IMPLIED
```

```
    xml:lang        NMTOKEN         #IMPLIED
    id              ID              #IMPLIED
    class           CDATA           #IMPLIED>
```

Elements That Might Contain This Element

ELEMENT	USED FOR
<fieldset>	Groups text and/or input fields
<p>	Paragraph of text
<pre>	Preformatted text

<setvar>

The <setvar> element is used to establish or reset the value of a WAP browser variable.

DTD Definition

```
<!ELEMENT setvar EMPTY>
<!ATTLIST setvar
    name            %vdata;         #REQUIRED
    value           %vdata;         #REQUIRED
    id              ID              #IMPLIED
    class           CDATA           #IMPLIED>
```

Elements That Might Contain This Element

ELEMENT	USED FOR
<go>	Invokes a URL
<prev>	Forces backward navigation
<refresh>	Refreshes the display

<small>

The <small> element is used to display text on a WAP browser in a font pitch that is smaller than normal.

DTD Definition

```
<!ELEMENT small (PCDATA | em | strong | b | i |
    u | big | small | br | img | anchor | a | table)*>
<!ATTLIST small
    xml:lang        NMTOKEN         #IMPLIED
```

```
    id              ID                  #IMPLIED
    class           CDATA               #IMPLIED>
```

Elements That Might Contain This Element

ELEMENT	USED FOR
<b>	Bold text
<big>	Larger-than-normal text
<em>	Emphasized text
<fieldset>	Groups text and/or input fields
<i>	Italic text
<p>	Paragraph of text
<small>	Smaller-than-normal text
<strong>	Stronger-than-normal text
<td>	Data element within a table
<u>	Underlined text

<strong>

The <strong> element is used to display text on a WAP browser in a font that is stronger than normal.

DTD Definition

```
<!ELEMENT strong (PCDATA | em | strong | b | i |
    u | big | small | br | img | anchor | a | table)*>
<!ATTLIST strong
    xml:lang        NMTOKEN             #IMPLIED
    id              ID                  #IMPLIED
    class           CDATA               #IMPLIED>
```

Elements That Might Contain This Element

ELEMENT	USED FOR
<b>	Bold text
<big>	Larger-than-normal text
<em>	Emphasized text
<fieldset>	Group text and/or input fields
<i>	Italic text
<p>	Paragraph of text

<small>	Smaller-than-normal text
<strong>	Stronger-than-normal text
<td>	Data element within a table
<u>	Underlined text

<table>

The <table> element is used to display text on a WAP browser in a tabular format.

DTD Definition

```
<!ELEMENT table (tr)+>
<!ATTLIST table
    title          CDATA          #IMPLIED
    align          CDATA          #IMPLIED
    columns        NMTOKEN        #REQUIRED
    xml:lang       NMTOKEN        #IMPLIED
    id             ID             #IMPLIED
    class          CDATA          #IMPLIED>
```

Elements That Might Contain This Element

ELEMENT	USED FOR
<b>	Bold text
<big>	Larger-than-normal text
<em>	Emphasized text
<fieldset>	Groups text and/or input fields
<i>	Italic text
<p>	Paragraph of text
<small>	Smaller-than-normal text
<strong>	Stronger-than-normal text
<u>	Underlined text

<td>

The <td> element is used to define table data within a <table> element.

DTD Definition

```
<!ELEMENT td (#PCDATA | em | strong | b | i | u |
    big | small | br | img | anchor | a )*>
```

```
<!ATTLIST td
    xml:lang        NMTOKEN             #IMPLIED
    id              ID                  #IMPLIED
    class           CDATA               #IMPLIED>
```

Elements That Might Contain This Element

ELEMENT	USED FOR
<tr>	Table row

The <tr> element is used to define a table row within a <table> element.

DTD Definition

```
<!ELEMENT tr (td)+>
<!ATTLIST tr
    id              ID                  #IMPLIED
    class           CDATA               #IMPLIED>
```

Elements That Might Contain This Element

ELEMENT	USED FOR
<table>	Formats data in rows and columns

<template>

The <template> element is used to define navigation elements that apply to all cards of a WML deck.

DTD Definition

```
<!ELEMENT template (do | onevent)*>
<!ATTLIST template
    onenterforward  %HREF;              #IMPLIED
    onenterbackward %HREF;              #IMPLIED
    ontimer         %HREF;              #IMPLIED
```

```
        id              ID                  #IMPLIED
        class           CDATA               #IMPLIED>
```

Elements That Might Contain This Element

ELEMENT	USED FOR
<wml>	Defines card-layer attributes that span all cards on a deck

<timer>

The <timer> element is used to start a timer on the browser.

DTD Definition

```
<!ELEMENT timer EMPTY>
<!ATTLIST timer
    name            NMTOKEN             #IMPLIED
    value           %vdata;             #REQUIRED
    id              ID                  #IMPLIED
    class           CDATA               #IMPLIED>
```

Elements That Might Contain This Element

ELEMENT	USED FOR
<card>	Defines a virtual page within a WML deck

<u>

The <u> element is used to underline display text.

DTD Definition

```
<!ELEMENT u (PCDATA | em | strong | b | i |
    u | big | small | br | img | anchor | a | table)*>
<!ATTLIST u
    xml:lang        NMTOKEN             #IMPLIED
    id              ID                  #IMPLIED
    class           CDATA               #IMPLIED>
```

Elements That Might Contain This Element

ELEMENT	USED FOR
<b>	Bold text
<big>	Larger-than-normal text
<em>	Emphasized text
<fieldset>	Groups text and/or input fields
<i>	Italic text
<p>	Paragraph of text
<small>	Smaller-than-normal text
<strong>	Stronger-than-normal text
<td>	Data element within a table
<u>	Underlined text

<wml>

The <wml> element is the root element of a WML deck.

DTD Definition

```
<!ELEMENT wml ( head?, template?, card+ )>
<!ATTLIST wml
    xml:lang        NMTOKEN                  #IMPLIED
    id              ID                       #IMPLIED
    class           CDATA                    #IMPLIED>
```

APPENDIX B

Java Servlet Development Kit (JSDK) 2.1.1 Reference

This appendix contains descriptions of the JSDK 2.1.1 Reference, as included with Sun Microsystems[3] Forte4j included on the accompanying CD-ROM in this book.

javax.servlet Package

The following sections look at the interface, classes, and exceptions defined within the `javax.servlet` package.

Interfaces

The following sections describe interfaces that are part of the `javax.servlet` package.

The RequestDispatcher Interface

```
public interface RequestDispatcher
{
  public void forward(final ServletRequest request,
    final ServletResponse response)
    throws ServletException, java,io.IOException;
  public void include(final ServletRequest request,
    final ServletResponse response)
```

```
        throws ServletException, java.io.IOException;
  }
```

The RequestDispatcher class can be used to dispatch messages to some other servlet server resource, such as other servlets. You might use this class if you were implementing an application than spans more than one servlet. Rather then redirecting from one servlet to another (a costly proposition on a wireless network), you can simply call the second servlet.

The forward( ) Method

```
public void forward(final ServletRequest request,
  final ServletResponse response)
  throws ServletException, java,io.IOException;
```

The forward() method is used to forward a request from one servlet to another servlet server resource. Calling this method implies that the receiving resource should handle the request.

The include( ) Method

```
public void include(final ServletRequest request,
  final ServletResponse response)
  throws ServletException, java.io.IOException;
```

The include() method is used to include content from another servlet server resource. Calling this method implies that the receiving resource should contribute to the processing of the request.

The Servlet Interface

```
public interface Servlet
{
  public void destroy();
  public ServletConfig getServletConfig()
  public String getServletInfo();
  public void init(final ServletConfig config)
    throws ServletException;
  public void service(final ServletRequest request,
    final ServletResponse response)
    throws ServletException, java.io.IOException;
}
```

All servlets must implement the Servlet interface. The HttpServlet class that we used in this book provides default implements of each method. The following sections describe the methods of the interface.

The destroy( ) Method

```
public void destroy();
```

The `destroy()` method provides the servlet a chance to clean up before the servlet engine terminates. As part of its shutdown procedure, a servlet engine will call the `destroy()` method for each servlet that was initialized during the execution of the engine. The default implementation does nothing.

The getServletConfig( ) Method

```
public ServletConfig getServletConfig();
```

The `getServletConfig()` method returns an object that implements the `Servlet-Config` interface. Servlets can use this object to query the initialization parameters for the servlet, as well as to obtain the `ServletContext`.

The getServletInfo( ) Method

```
public String getServletInfo();
```

The `getServletInfo()` method is used to obtain the information string associated with the servlet. Your servlet implementation should return a string that contains a readable text string, including information such as the name of the servlet, the author, and/or the copyright notice.

The init( ) Method

```
public void init(final ServletConfig config)
  throws ServletException;
```

This `init()` method provides the servlet a chance to initialize itself before handling the first service request. Many servlet engines provide a means for this method to be called at startup time. Otherwise (and by default), the `init()` method will be called immediately prior to and as a result of receiving its first service request. The `config` argument is an instance of an object that implements the `ServletConfig` interface. This object can be used to obtain the initialization parameters for the servlet.

The default implementation of the `init()` method invokes the `log()` method of the servlet to indicate that the servlet has been initialized. When overloading this method, you should call the superclass's implementation of `init()`.

The service( ) Method

```
public void service(ServletRequest request,
  ServletResponse response)
  throws ServletException, java.io.IOException;
```

The `service()` method is used to dispatch the request according to what is appropriate for the protocol (scheme) of the servlet. For example, in the case of the HyperText Transport Protocol (HTTP), the default implementation of this method will call the appropriate HTTP method handler based on the HTTP method of the request (*get*, *post*, *delete*, and so on).

The ServletConfig Interface

```
public interface ServletConfig
{
  public String getInitParameter(String strName);
  public java.util.Enumeration getInitParameterNames();
  public ServletContext getServletContext();
}
```

All servlet engines need to maintain a record of configuration information for the servlets that they load. This interface defines the methods expected by servlets. The HTTP servlet engine provided on the accompanying CD-ROM holds this information in the class `ServletConfigImp`. Unless you are writing your own servlet engine or extending mine, you will probably not need to implement this interface.

The getInitParameter( ) Method

```
public String getInitParameter(String strName);
```

This method returns one of the servlet's initialization parameters by name.

The getInitParameterNames( ) Method

```
public java.util.Enumeration getInitParameterNames();
```

This method returns all the names of a servlet's initialization parameters.

The getServletContext( ) Method

```
public ServletContext getServletContext();
```

Servlets use this method to obtain context information. Some servlet engines (such as Apache) implement multiple contexts. The HTTP servlet engine provided on the accompanying CD-ROM implements a single context by using the class `ServletContextImp`. Unless you are writing your own servlet engine or perhaps extending mine to support more than one context, you will probably not need to implement this interface.

The ServletContext Interface

```
public interface ServletContext
{
  public Object getAttribute(String strName);
```

```
    public java.util.Enumeration getAttributeName();
    public ServletContext getContext(String strName);
    public int getMajorVersion();
    public String getMimeType(String strExtension);
    public int getMinorVersion();
    public String getRealPath(String strRelativePath);
    public RequestDispatcher getRequestDispatcher(String strName);
    public java.net.URL getResource(String strURL)
      throws java.net.MalformedURLException;
    public java.io.InputStream getResourceAsStream(String strURL)
      throws java.net.MalformedURLException;
    public String getServerInfo();
    public Servlet getServlet(String strName)
    public throws ServletException;
    public java.util.Enumeration getServletNames();
    public java.util.Enumeration getServlets();
    public void log(String strMessage, final Throwable throwable);
    public void log(final Throwable throwable);
    public void log(String strMessage);
    public void removeAttribute(String strName);
    public void setAttribute(String strName,
      final Object objAttribute);
}
```

The `ServletContext` interface contains information about the environment within which the servlet runs. Some servlet engines implement the notion of multiple virtual environments. The `ServletContext` object contains information about the virtual environment at hand. The servlet engine on the accompanying CD-ROM implements only a single `ServletContext`.

The getAttribute( ) Method

```
public Object getAttribute(String strName);
```

The `getAttribute()` method returns an `Object` associated with the `ServletContext` by name. This capability enables servlets belonging to the same context to pass data.

The getAttributeName( ) Method

```
public java.util.Enumeration getAttributeName();
```

The `getAttributeName()` method returns the name of all name-value pair attributes associated with the `ServletContext`.

The getContext( ) Method

```
public ServletContext getContext(String strName);
```

The `getContext()` method returns the context associated with a particular URI path. If your servlet needs access to the `ServletContext` of another servlet engine resource, this method can be used to find that `ServletContext`.

The getMajorVersion( ) Method

```
public int getMajorVersion();
```

This method returns the major version number of the JSDK supported by the servlet engine within the context at hand. For example, version 2.1.1 of the JSDK would return 2.

The getMimeType( ) Method

```
public String getMimeType(String strFile);
```

This method will return the Multi-Purpose Internet Mail Extensions (MIME) type of the filename passed. If the MIME type is not known, a null is returned.

The getMinorVersion( ) Method

```
public int getMinorVersion();
```

This method returns the minor version number of the JSDK supported by the servlet engine within the context at hand. For example, version 2.1.1 of the JSDK would return 1.

The getRealPath( ) Method

```
public String getRealPath(String strURIPath);
```

The `getRealPath()` method returns the real absolute path of a file passed by its URI path.

The getRequestDispatcher( ) Method

```
public RequestDispatcher getRequestDispatcher(String strName);
```

The `getRequestedDispatcher()` method returns the `RequestDispatcher` object associated with the URI path argument.

The getResource( ) Method

```
public java.net.URL getResource(String strURL)
    throws java.net.MalformedURLException;
```

This method returns a URL object representing the URL specification argument. If the URL is well formed yet does not point to a valid resource, a null is returned.

The getResourceAsStream( ) Method

```
public java.io.InputStream getResourceAsStream(String strURL)
    throws java.net.MalformedURLException;
```

This method returns an `InputStream` object for the specified URL argument. If the URL is well formed yet does not point to a valid resource, a null is returned.

The getServerInfo( ) Method

```
public String getServerInfo();
```

The `getServerInfo()` method returns a readable `String` that describes the servlet engine.

The getServlet( ) Method

```
public Servlet getServlet(String strName)
  throws ServletException;
```

This method returns a reference to a servlet by name. If the servlet cannot be located (or has not yet been initialized), a null is returned.

The getServletNames( ) Method

```
public java.util.Enumeration getServletNames();
```

The `getServletNames()` method returns an `Enumeration` of servlet names that are known to the context.

The getServlets( ) Method

```
public java.util.Enumeration getServlets();
```

The `getServlets()` method returns an `Enumeration` of servlets that are known to the context.

The log( ) Method

```
public void log(String strMessage, final Throwable throwable);
public void log(final Throwable throwable);
public void log(String strMessage);
```

The `log()` method writes a message and/or a stack trace to the log associated with the context. Some servlet engines maintain separate logs on a per-context basis.

The removeAttribute( ) Method

```
public void removeAttribute(String strName);
```

The `removeAttribute()` method removes the named attribute from the context attribute pool.

The setAttribute( ) Method

```
public void setAttribute(String strName, final Object objAttribute);
```

The `setAttribute()` method sets or establishes a name-value pair attribute within the context's attribute pool.

The ServletRequest Interface

```
public interface ServletRequest
{
  public Object getAttribute(String name);
  public java.util.Enumeration getAttributeNames();
  public String getCharacterEncoding();
  public int getContentLength();
  public String getContentType();
  public ServletInputStream getInputStream()
    throws IOException;
  public String getParameter(String name);
  public Enumeration getParameterNames();
  public String[] getParameterValues(String name);
  public String getProtocol();
  public BufferedReader getReader ()
    throws IOException;
  public String getRealPath(String path);
  public String getRemoteAddr();
  public String getRemoteHost();
  public String getScheme();
  public String getServerName();
  public int getServerPort();
  public void setAttribute(String strName, final Object objValue);
}
```

Servlets use an object that implements the `ServletRequest` interface to obtain URL and protocol-header information. The `HttpServletRequest` class implements this interface. The HTTP servlet engine on the accompanying CD-ROM uses the `HttpServletRequestImp` class to represent an `HttpServletRequest`. This class implements the `ServletRequest` interface. As a developer of WAP Servlets, you will most likely not need to implement this interface.

The getAttribute( ) Method

```
public Object getAttribute(String name);
```

This method returns an HTTP header attribute by name. This method is useful for getting attributes that are not covered by another method.

The getAttributeNames( ) Method

```
public java.util.Enumeration getAttributeNames();
```

This method returns the names of the HTTP header name-value pairs as an `Enumeration`. If for some reason no attributes are present, an empty `Enumeration` is returned.

The getCharacterEncoding( ) Method

```
public BufferedReader getCharacterEncoding();
```

This method returns the character set encoding of the request.

The getContentLength( ) Method

```
public int getContentLength();
```

This method returns the length (in bytes) of the request's content, and if the length of the content is unknown, −1 is returned. This value is the same as the Common Gateway Interface (CGI) variable `CONTENT_LENGTH`.

The getContentType( ) Method

```
public String getContentType();
```

This method returns the MIME type of the request's content. If the MIME type is unknown or if there is no content associated with the request, a null is returned. This value is the same as the CGI variable `CONTENT_TYPE`.

The getInputStream( ) Method

```
public ServletInputStream getInputStream()
  throws IOException;
```

This method returns an input stream for reading binary data from the content portion of the request. For each request, a developer can only request an input stream via this method or a print reader via the method `getReader()`. If the developer calls this method after calling `getReader()`, an `IllegalStateException` is thrown.

The getParameter( ) Method

```
public String getParameter(String strName);
```

This method returns the value of a request parameter by name (`strName`) as a string. A parameter is defined as a name-value pair either within the URL or within the content of the request. If the requested parameter does not exist in the request, a null is returned. If developers expect that multiple values will be associated with a given parameter name, they should use the `getParameterValues()` method.

The getParameterNames( ) Method

```
public Enumeration getParameterNames();
```

This method returns all parameter names associated with the request as an `Enumeration`. A parameter is defined as a name-value pair either within the URL or within the content of the request.

The getParameterValues( ) Method

```
public String[] getParameterValues(String strName);
```

This method returns the multiple values of a request parameter by name (`strName`) as an array of strings. A parameter is defined as a name-value pair either within the URL or within the content of the request. If the requested parameter does not exist in the request, a null is returned.

The getProtocol( ) Method

```
public String getProtocol();
```

This method returns the protocol and version of the request. The format of this value is as follows: `<protocol>/<major version>. <minor version>`. This value is the same as the CGI variable `SERVER_PROTOCOL`.

The getReader( ) Method

```
public BufferedReader getReader()
  throws IOException;
```

This method returns a reader for reading character data from the content portion of the request. For each request, developers can only request a reader via this method or an input stream via the method `getInputStream()`. If the developer calls this method after calling `getInputStream()`, an `IllegalStateException` is thrown.

The getRealPath( ) Method

```
public String getRealPath(String strPath);
```

Given a relative path to the document root of the server (`strPath`), this method returns the actual absolute path to the given resource.

The getRemoteAddr( ) Method

```
public int getRemoteAddr();
```

This method returns the IP address of the machine that sent the request (in other words, the machine running the browser that initiated the request). This value is the same as the CGI variable `REMOTE_ADDR`.

The getRemoteHost( ) Method

```
public int getRemoteHost();
```

This method returns the Domain Name Service (DNS) of the machine that sent the request. If the name cannot be resolved, this value will contain the Internet Protocol

(IP) address of the machine as a string. This value is the same as the CGI variable REMOTE_HOST.

The getScheme( ) Method

```
public String getScheme();
```

This method returns the scheme (or protocol) of the request's URL (for example, http, https, or ftp). Each scheme can have its own rules for URL construction. The format of standard URLs by scheme can be found in RFC 1738.

The getServerName( ) Method

```
public String getServerName();
```

This method returns the DNS host name of the server that received the request (the server that is executing the servlet at hand). This value is the same as the CGI variable SERVER_NAME.

The getServerPort( ) Method

```
public int getServerPort();
```

This method returns the TCP port number on which the request was received. This value is the same as the CGI variable SERVER_PORT.

The setAttribute( ) Method

```
public void setAttribute(String strName, final Object objValue);
```

The `setAttribute()` method can be used to establish or change the value of an HTTP header name-value pair.

The ServletResponse Interface

```
public interface ServletResponse
{
    public String getCharacterEncoding();
    public ServletOutputStream getOutputStream()
        throws IOException;
    public PrintWriter getWriter()
        throws IOException;
    public void setContentLength(int nLength);
    public void setContentType(String strType);
}
```

Servlets use an object that implements the `ServletReponse` interface in order to respond to a request. The `HttpServletResponse` class implements this interface. The HTTP servlet engine on the accompanying CD-ROM uses the `HttpServletResponseImp` class to represent an `HttpServletResponse`. This class implements the

ServletResponse interface. As a developer of WAP Servlets, you will most likely not need to implement this interface.

The getCharacterEncoding( ) Method

```
public String getCharacterEncoding();
```

By definition, the content portion of an HTTP packet is MIME encoded. This method returns the character encoding used to encode the content of this response. The default character encoding is ISO-8859-1.

The getOutputStream( ) Method

```
public ServletOutputStream getOutputStream()
  throws IOException;
```

This method returns an output stream for writing binary data to the content portion of the response. For each response, a developer can only request an output stream via this method or a print writer via the method getWriter(). If the developer calls this method after calling getWriter(), an IllegalStateException is thrown.

The getWriter( ) Method

```
public PrintWriter getWriter()
  throws IOException;
```

This method returns a print writer for writing text data to the content portion of the response. For each response, a developer can only request an print writer via this method or a output stream via the method getOutputStream(). If the developer calls this method after calling getOutputStream(), an IllegalStateException is thrown.

The setContentLength( ) Method

```
public void setContentLength(int nLength);
```

This method is used to establish the length of the content section of the HTTP response (the content-length HTTP header entry). Browsers use this value to aid in their processing of a packet.

The setContentType( ) Method

```
public abstract void setContentType(String strType);
```

This method is used to establish the MIME type of the content section of the HTTP response (the content-type HTTP header entry). Browsers use this value to understand how to process the content section.

The SingleThreadModel Interface

```
public interface SingleThreadModel
{
}
```

Normally, a servlet engine will allow multiple threads to pass through a servlet at the same time. Servlets that implement the `SingleThreadModel` interface are guaranteed to be single threaded. While in rare instances this behavior might be desirable, it obviously causes a bottleneck with servlet requests backing up on the server while they wait their turn.

Classes

The following sections describe classes that are part of the `javax.servlet` package.

The GenericSevlet Class

```
public abstract GenericServlet extends Object
{
  public GenericServlet();
  public void destroy();
  public String getInitParameter(String strName);
  public java.lang.Enumeration getInitParameterNames();
  public ServletConfig getServletConfig();
  public ServletContext getServletContext();
  public String getServletInfo();
  public void init(final ServletConfig config)
    throws ServletException;
  public void init()
    throws ServletException;
  public void log(String strMessage);
  public void log(String strMessage, final Throwable throwable);
  public void service(ServletRequest request,
    ServletResponse response)
    throws ServletException, java.io.IOException;
}
```

The `GenericServlet` class is the superclass of the `HttpServlet`. The following sections describe the methods of this class.

The GenericServlet( ) Method

```
public GenericServlet();
```

This method constructs a new `GenericServlet`.

The destroy( ) Method

```
public void destroy();
```

The `destroy()` method provides a default implementation required by the `Servlet` interface. The default implementation does nothing.

The getInitParameter( ) Method

```
public String getInitParameter(String strName);
```

This method returns one of the servlet's initialization parameters by name.

The getInitParameterNames( ) Method

```
public java.lang.Enumeration getInitParameterNames();
```

This method returns all of the names of a servlet's initialization parameters.

The getServletConfig( ) Method

```
public ServletConfig getServletConfig();
```

The `getServletConfig()` method returns an object that implements the `ServletConfig` interface. Servlets can use this object to query the initialization parameters for the servlet as well as to obtain the `ServletContext`.

The getServletContext( ) Method

```
public ServletContext getServletContext();
```

The `getServletContext()` method is essentially a convenience method for retrieving the `ServletContext` (refer to the section regarding `ServletConfig.getServletContext()`).

The getServletInfo( ) Method

```
public String getServletInfo();
```

The `getServletInfo()` method is used to obtain the information string associated with the servlet. Your servlet implementation should return a string that contains a readable text string, including information such as the name of the servlet, the author, and/or the copyright notice.

The init( ) Method

```
public void init(final ServletConfig config)
  throws ServletException;

public void init()
  throws ServletException;
```

This init() method provides the servlet a chance to initialize itself before handling the first service request (refer to Servlet.init()).

The log( ) Method

```
public void log(String strMessage);

public void log(String strMessage, final Throwable throwable);
```

The log() method provides a servlet with a means of writing a message to the servlet engine's log. Some servlet engines maintain separate logs for each ServletContext. The version that takes a Throwable will print a stack trace, as well as the message, to the log.

The service( ) Method

```
public abstract void service(ServletRequest request,
  ServletResponse response)
  throws ServletException, java.io.IOException;
```

The service() method is used to dispatch the request according to what is appropriate for the protocol (scheme) of the servlet. For example, in the case of HTTP, the default implementation of this method will call the appropriate HTTP method handler based on the HTTP method of the request (*get*, *post*, *delete*, and so on).

The ServletInputStream Class

```
public abstract class ServletInputStream extends java.io.InputStream
{
  protected ServletInputStream();
  public int readLine(byte [] buffer, int nOffset, int nLength)
    throws java.io.IOException;
}
```

A ServletInputStream is returned by Servlet.getInputStream() and can be used to read the content portion of a request.

The readLine( ) Method

```
public int readLine(byte [] buffer, int nOffset, int nLength)
  throws java.io.IOException;
```

The readLine() method is used to read nLength bytes of data into the buffer starting at index nOffset. If a new line (\n) is encountered, the read is terminated. The actual number of bytes read is returned.

The ServletOutputStream Class

```
public abstract class ServletOutputStream extends OutputStream
{
```

```
    protected ServletOutputStream();
    public void print(int nValue)
      throws java.io.IOException;
    public void print(long lValue)
      throws java.io.IOException;
    public void print(float fValue)
      throws java.io.IOException;
    public void print(boolean bValue)
      throws java.io.IOException;
    public void print(double dValue)
      throws java.io.IOException;
    public void print(String strValue)
      throws java.io.IOException;
    public void print(char c)
      throws java.io.IOException;
    public void println(char c)
      throws java.io.IOException;
    public void println(String strValue)
      throws java.io.IOException;
    public void println(int nValue)
      throws java.io.IOException;
    public void println(long lValue)
      throws java.io.IOException;
    public void println()
      throws java.io.IOException;
    public void println(boolean bValue)
      throws java.io.IOException;
    public void println(double dValue)
      throws java.io.IOException;
    public void println(float fValue)
      throws java.io.IOException;
  }
```

A `ServletOutputStream` is returned by the `ServletResponse.getOutputStream()` method. This method provides several print functions for writing information to the content portion of a response.

The print( ) Method

```
public void print(int nValue)
  throws java.io.IOException;

public void print(long lValue)
  throws java.io.IOException;

public void print(float fValue)
  throws java.io.IOException;

public void print(boolean bValue)
  throws java.io.IOException;
```

```
public void print(double dValue)
  throws java.io.IOException;

public void print(String strValue)
  throws java.io.IOException;

public void print(char c)
  throws java.io.IOException;
```

A variety of standard `print()` methods are provided for writing to the `ServletOutputStream` at hand.

The println( ) Method

```
public void println(char c)
  throws java.io.IOException;

public void println(String strValue)
  throws java.io.IOException;

public void println(int nValue)
  throws java.io.IOException;

public void println(long lValue)
  throws java.io.IOException;

public void println()
  throws java.io.IOException;

public void println(boolean bValue)
  throws java.io.IOException;

public void println(double dValue)
  throws java.io.IOException;

public void println(float fValue)
  throws java.io.IOException;
```

A variety of standard `println()` methods are provided for writing to the `ServletOutputStream` at hand.

Exceptions

The following sections describe exceptions that are part of the `javax.servlet` package.

The ServletException Exception Class

```
public class ServletException extends Exception
{
```

```
  public ServletException(java.lang.Throwable throwable);
  public ServletException(String strMessage);
  public ServletException(String strMessage,
    java.lang.Throwable throwable);
  public ServletException();
  public java.lang.Throwable getRootCause();
}
```

The `ServletException` is provided for throwing exceptions from within a servlet.

The ServletException( ) Constructor

```
public ServletException(java.lang.Throwable throwable);
public ServletException(String strMessage);
public ServletException(String strMessage,
  java.lang.Throwable throwable);
public ServletException();
```

This constructor builds a `ServletException` object based on a message and/or a root cause (a throwable).

The getRootCause( ) Method

```
public java.lang.Throwable getRootCause();
```

This method enables a method that catches a `ServletException` to discover the root cause of the exception.

The UnavailableException Exception Class

```
public UnavailableException
{
  public UnavailableException(Servlet servlet, String strMessage);
  public UnavailableException(int nSeconds, Servlet servlet,
    String strMessage);
  public boolean isPermanent();
  public Servlet getServlet();
  public int getUnavailableSeconds();
}
```

The `UnavailableException` can be thrown by a servlet if it is either temporarily or permanently unable to process requests.

The UnavailableException( ) Constructor

```
public UnavailableException(Servlet servlet, String strMessage);
```

This form of the constructor is used to indicate a permanently unavailable servlet.

```
public UnavailableException(int nSeconds, Servlet servlet,
  String strMessage);
```

This form of the constructor is used to indicate a temporary condition. The nSeconds argument should contain the estimated number of seconds that the servlet will not be available.

The isPermanent( ) Method

```
public boolean isPermanent();
```

This method returns *true* if a permanent condition was reported by the servlet. Otherwise, *false* is returned.

The getServlet( ) Method

```
public Servlet getServlet();
```

This method returns the Servlet argument passed to the constructor.

The getUnavailableSeconds( ) Method

```
public int getUnavailableSeconds();
```

This method returns the nSeconds argument passed to the constructor.

javax.servlet.http Package

The following sections look at the interfaces and classes defined within the javax.servlet.http package.

Interfaces

The following sections describe interfaces that are part of the javax.servlet.http package.

The HttpServletRequest Interface

```
public interface HttpServletRequest
extends ServletRequest
{
  public String getAuthType();
  public Cookies [] getCookies();
  public long getDateHeader(String name);
  public String getHeader(String strName);
  public Enumeration getHeaderNames();
  public int getIntHeader(String strName);
  public String getMethod();
  public String getPathInfo();
  public String getPathTranslated();
```

```
    public String getQueryString();
    public String getRemoteUser();
    public String getRequestedSessionID();
    public String getRequestURI();
    public String getServletPath();
    public HttpSession getSession();
    public HttpSession getSession(boolean bCreate);
    public boolean isRequestedSessionFromCookie();
    public boolean isRequestedSessionFromURL();
    public boolean isRequestedSessionFromUrl();
    public boolean isRequestedSessionValid();
}
```

An instance of a class implementing the `HttpServletRequest` interface is passed to the servlet as a representative of a request that invoked an `HttpServlet`. This object provides convenience methods for accessing the various attributes of an HTTP request. The following sections describe the functionality of the methods.

The getAuthType( ) Method

```
public String getAuthType();
```

The `getAuthType()` method returns information that describes the authentication type of the HTTP request.

The getCookies( ) Method

```
public Cookies [] getCookies();
```

The `getCookies()` method returns all cookies specified in the HTTP header. If no cookies were passed in the request, a *null* is returned by this method.

The getDateHeader( ) Method

```
public long getDateHeader(String name);
```

The `getDateHeader()` method returns the named parameter as a `long` value. This value is suitable for passing to java.util.Date in order to generate a date value. You should note that this method should only be invoked for name-value pairs that represent a date. If there is no name-value pair with the specified name, this method returns a zero.

The getHeader( ) Method

```
public String getHeader(String strName);
```

The `getHeader()` method returns the named parameter as a `String` value. This method can be called for any name-value pair in the HTTP header. If there is no name-value pair with the specified name, a *null* is returned by this method.

The getHeaderNames( ) Method

```
public Enumeration getHeaderNames();
```

This method returns an `Enumeration` of all HTTP header name-value pairs by name. This method is useful for displaying all name-value pairs for debugging purposes.

The getIntHeader( ) Method

```
public int getIntHeader(String strName);
```

The `getIntHeader()` method returns the named parameter as an `int` value. You should note that this method should only be invoked for name-value pairs that have an integer value. If there is no name-value pair with the specified name, this method returns a zero.

The getMethod( ) Method

```
public String getMethod();
```

The `getMethod()` method returns the method specified on line 1 of the HTTP request. The value returned is one of the HTTP access methods: *delete, get, options, post, put,* or *trace.*

The getPathInfo( ) Method

```
public String getPathInfo();
```

The `getPathInfo()` method returns the path portion of the URI specified on line 1 of the HTTP request.

The getPathTranslated( ) Method

```
public String getPathTranslated();
```

The `getPathTranslated()` method returns the real absolute path to the resource specified on line 1 of the HTTP request.

The getQueryString( ) Method

```
public String getQueryString();
```

The `getQueryString()` method returns the query string portion of the URI specified on line 1 of the HTTP request.

The getRemoteUser( ) Method

```
public String getRemoteUser();
```

The `getRemoteUser()` method returns the username specified in the HTTP header.

The getRequestedSessionID( ) Method

```
public String getRequestedSessionID();
```

This method returns the session ID specified in either the query string or the session cookie.

The getRequestURI( ) Method

```
public String getRequestURI();
```

This method returns the URI specified on line 1 of the HTTP request.

The getServletPath( ) Method

```
public String getServletPath();
```

The `getServletPath()` method returns the path information following the logical path to the servlet on the first line of the HTTP request.

The getSession( ) Method

```
public HttpSession getSession();
public HttpSession getSession(boolean bCreate);
```

The `getSession()` method is used to look up and/or create the session structure associated with the request.

The isRequestSessionFromCookie( ) Method

```
public boolean isRequestedSessionFromCookie();
```

This method returns *true* if the requested session ID came from a cookie. Otherwise, *false* is returned.

The isRequestedSessionFromURL( ) Method

```
public boolean isRequestedSessionFromURL();
public boolean isRequestedSessionFromUrl();
```

This method returns *true* if the requested session ID came from the query string portion of the URL. Otherwise, *false* is returned. Both forms of this method perform the same function.

The isRequestedSessionValid( ) Method

```
public boolean isRequestedSessionValid();
```

This method returns *true* if the requested session value (either requested in the query string or in a cookie) represents a currently valid session. Otherwise, *false* is returned.

The HttpServletResponse Interface

```
public interface HttpServletResponse
  extends ServletResponse
{
  public void addCookie(Cookie cookie);
  public boolean containsHeader(String strName);
  public String encodeRedirectURL(String strURL);
  public String encodeRedirectUrl(String strURL);
  public String encodeURL(String strURL);
  public String encodeUrl(String strURL);
  public void sendError(int nErrorCode, String strMessage)
    throws java.io.IOException;
  public void sendError(int nErrorCode)
    throws java.io.IOException;
  public void sendRedirect(String strURL)
    throws java.io.IOException;
  public void setDateHeader(String strName, long lDateValue);
  public void setHeader(String strName, String strValue);
  public void setIntHeader(String strName, int nValue);
  public void setStatus(int nErrorCode);
  public void setStatus(int nErrorCode, String strMessage);
}
```

An instance of a class implementing the `HttpServletResponse` interface is passed to the servlet as a representative of a would-be response to the `HttpServletRequest` associated with the servlet invocation. This object provides convenience methods for accessing the various attributes of an HTTP response. The following sections describe the functionality of the methods.

The addCookie( ) Method

```
public void addCookie(Cookie cookie);
```

The `addCookie()` method is used to add a cookie specification to the response.

The containsHeader( ) Method

```
public boolean containsHeader(String strName);
```

This method returns true if the specified HTTP header name-value pair has been established in the response.

The encodeURL( ) Method

```
public String encodeURL(String strURL);
public String encodeUrl(String strURL);
```

The `encodeURL()` method is used to encode the session ID in the query string portion of a specified URL. If the servlet knows that the requesting client supports (and accepts) cookies, the session ID is specified as a cookie automatically, and the string returned by this method will be the same as the strURL argument. If the client does not support cookies, however, a name-value pair is added to the query string of the strURL argument to form the URL that is returned. If the cookie-support capabilities of the client are not yet known, this method will encode the URL into the return value.

This method should be used when a servlet renders a URL (for callback purposes) associated with the session at hand.

The encodeRedirectURL( ) Method

```
public String encodeRedirectURL(String strURL);
public String encodeRedirectUrl(String strURL);
```

The encodeRedirectURL() method is used to encode the session ID in the query string portion of a specified URL. If the servlet knows that the requesting client supports (and accepts) cookies, the session ID is specified as a cookie automatically, and the string returned by this method will be the same as the strURL argument. If the client does not support cookies, however, a name-value pair is added to the query string of the strURL argument to form the redirect URL that is returned. If the cookie-support capabilities of the client are not yet known, this method will encode the URL into the return value.

This method should be used when a servlet renders a redirect URL associated with the session at hand.

The sendError( ) Method

```
public void sendError(int nErrorCode, String strMessage)
  throws java.io.IOException;

public void sendError(int nErrorCode)
  throws java.io.IOException;
```

Normally, an error code of 200 (OK) and a message of "OK" are returned by a servlet. If an error condition exists, however, the sendError() method can be used to report the error along with an option message that further describes the error condition in readable text.

Unlike the setStatus() method, the sendError() method sends a response with an empty content immediately. If you want to report an error but still provide content in your response, you should use the setStatus() method.

The sendRedirect( ) Method

```
public void sendRedirect(String strURL)
  throws java.io.IOException;
```

The sendRedirect() method is used to send a redirect response to the client. A redirect instructs the client to immediately make a request to an alternate URL.

The setDateHeader( ) Method

```
public void setDateHeader(String strName, long lDateValue);
```

The `setDateHeader()` method is used to establish an HTTP header name-value pair that represents a date. Given the `lDateValue` argument that represents a date, this method will generate a date/time value in the appropriate ASCII format.

The setHeader( ) Method

```
public void setHeader(String strName, String strValue);
```

The `setHeader()` method is used to establish an HTTP header name-value pair that represents a string value. This method can be used to establish any HTTP header option.

The setIntHeader( ) Method

```
public void setIntHeader(String strName, int nValue);
```

The `setIntHeader()` method is used to establish an HTTP header name-value pair that represents an integer value.

The setStatus( ) Method

```
public void setStatus(int nErrorCode);
public void setStatus(int nErrorCode, String strMessage);
```

The `setStatus()` method is used to establish an error code and message other than the default (error code 200 and message "OK").

Use the `setStatus()` method if you are still returning content in your response; otherwise, use the `sendError()` method.

The HttpSession Interface

```
public interface HttpSession
{
  public long getCreationTime();
  public String getId();
  public long getLastAccessedTime();
  public int getMaxInactiveInterval();
  public HttpSessionContext getSessionContext();
  public Object getValue(String strName);
  public String [] getValueNames();
  public void invalidate();
  public boolean isNew();
  public void putValue(String strName, Object objValue);
  public void removeValue(String strName);
```

```
    public void setMaxInactiveInterval(int nInterval);
  }
```

An instance of a class implementing the HttpSession interface is returned by the method HttpRequest.getSession(). The session object is a vehicle for persistent storage of user information across multiple invocations of a servlet. Because the HTTP protocol is stateless, the HttpSession provides a means for users to have a stateful session. The following sections describe the methods that are available to the HttpSession.

The getCreationTime( ) Method

```
public long getCreationTime();
```

This method returns the date/time that a session object was originally created.

The getId( ) Method

```
public String getId();
```

The getId() method returns the session ID of the session at hand.

The getLastAccessedTime( ) Method

```
public long getLastAccessedTime();
```

The getLastAccessTime() returns the date/time that a session object was last accessed.

The getMaxInactiveInterval( ) Method

```
public int getMaxInactiveInterval();
```

The getMaxInactivityInterval() method returns the maximum number of seconds that a servlet session can be inactive before it is invalidated.

The getSessionContext( ) Method

```
public HttpSessionContext getSessionContext();
```

This method returns the context of the session.

The getValue( ) Method

```
public Object getValue(String strName);
```

The getValue() method returns a value previously established with respect to the session. Session values are used to store persistent information that is associated with a servlet session.

The getValueNames( ) Method

```
public String [] getValueNames();
```

The `getValueNames()` method returns an array of value names associated with the session. Session values are used to store persistent information that is associated with a servlet session. This method is useful for displaying the name and value of all values associated with a session for debugging purposes.

The invalidate( ) Method

```
public void invalidate();
```

This method makes a session invalid. A session is not useable once it has been invalidated. Invalidating a session is likened to user logoff.

The isNew( ) Method

```
public boolean isNew();
```

The `isNew()` method returns *true* during the request in which it was created. Otherwise, *false* is returned.

The putValue( ) Method

```
public void putValue(String strName, Object objValue);
```

The `putValue()` method sets a value for the balance of the session, and session values are used to store persistent information that is associated with a servlet session.

The removeValue( ) Method

```
public void removeValue(String strName);
```

The `removeValue()` method is used to remove an existing session value.

The setMaxInactiveInterval( ) Method

```
public void setMaxInactiveInterval(int nInterval);
```

The `setMaxInactiveInterval()` is used to establish the maximum number of seconds that a session will remain active. Once the maximum inactive interval has elapsed, the servlet engine will invalidate the session.

The HttpSessionBindingListener Interface

```
public interface HttpSessionBindingListener
  extends java.util.EventListener
{
  public void valueBound(final HttpSessionBindingEvent event);
```

```
    public void valueUnbound(final HttpSessionBindingEvent event);
}
```

A class can become a session-binding listener by implementing the `HttpSessionBindingListener` interface.

The valueBound( ) Method

```
public void valueBound(final HttpSessionBindingEvent event);
```

This method is called in order to notify the listener that a value has been bound.

The valueUnbound( ) Method

```
public void valueUnbound(final HttpSessionBindingEvent event);
```

This method is called in order to notify the listener that a value has been unbound.

The HttpSessionContext Interface

```
public interface HttpSessionContext
{
    public java.util.Enumeration getIds();
    public HttpSession getSession(String strName);
}
```

The `HttpSessionContext` interface provides methods for accessing sessions that are associated with a servlet. The following sections describe those methods.

The getIds( ) Method

```
public java.util.Enumeration getIds();
```

The `getIds()` method returns an `Enumeration` of session IDs associated with a servlet. This method is useful for servlets that want to perform an action relative to their collective servlets.

The getSession( ) Method

```
public HttpSession getSession(String strID);
```

Given a session ID, the `getSession()` method can be used to look up the session that is associated with the ID.

Classes

The following sections describe classes that are part of the `javax.servlet.http` package.

Cookie Class

```
public class Cookie extends Object Implements Clonable
{
  public Cookie (String strName, String strValue);
  public Object clone();
  public String getComment();
  public String getDomain();
  public int getMaxAge();
  public String getName();
  public String getPath();
  public boolean getSecure();
  public String getValue();
  public int getVersion();
  public void setComment(String strPurpose);
  public void setDomain(String strDomain);
  public void setMaxAge(int nExpiry);
  public void setPath(String strURL);
  public void setSecure(boolean bSecure);
  public void setValue(String newValue);
  public void setVersion(int nVersion);
}
```

The `Cookie` class is used to access the properties of an HTTP cookie. The following sections describe these methods.

The Cookie( ) Method

```
public Cookie (String strName, String strValue);
```

This method constructs a `Cookie` object with an initial name of `strName` and an initial value of `strValue`.

The clone( ) Method

```
public Object clone();
```

This method returns a new `Cookie` object that is an exact copy of this cookie.

The getComment( ) Method

```
public String getComment();
```

This method returns the comment associated with this cookie.

The getDomain( ) Method

```
public String getDomain();
```

This method returns the domain associated with this cookie.

The getMaxAge( ) Method

```
public int getMaxAge();
```

This method returns the number of seconds before this cookie will expire.

The getName( ) Method

```
public String getName();
```

This method returns the name of this cookie.

The getPath( ) Method

```
public String getPath();
```

This method returns the path associated with this cookie.

The getSecure( ) Method

```
public boolean getSecure();
```

This method returns true if this cookie requires a secure connection; otherwise, false is returned.

The getValue( ) Method

```
public String getValue();
```

This method returns the value associated with this cookie.

The getVersion( ) Method

```
public int getVersion();
```

This method returns the version of this cookie.

The setComment( ) Method

```
public void setComment(String strPurpose);
```

This method establishes a comment based on the value of `strPurpose` for this cookie. The default comment of a cookie is null.

The setDomain( ) Method

```
public void setDomain(String strPattern);
```

This method establishes a domain-name pattern (based on `strPattern`) for servers that are to receive this cookie. By default, cookies are only sent to the server that created them. If other servers require the use of this cookie, the developer should establish a domain-name pattern that includes all such servers. The path and security flag are also taken into consideration when sending a cookie.

The setMaxAge( ) Method

```
public void setMaxAge(int nSeconds);
```

This method establishes the number of seconds (based on nSeconds) that the cookie will live.

The setPath( ) Method

```
public void setPath(String strURL);
```

This method establishes a path (based on strURL) for the cookie, and once stored, the browser will send the cookie when making requests to URLs at or below the established path. The domain and security flag are also taken into consideration when sending a cookie.

The setSecure( ) Method

```
public void setSecure(boolean bSecure);
```

This method establishes the value (based on bSecure) of the security flag. If bSecure is true, this cookie will only be sent over a secure connection. Otherwise, the cookie will be sent according to its path and domain. The default value for the security flag is *false*. The domain and path are also taken into consideration when sending a cookie.

The setValue( ) Method

```
public void setValue(String strNewValue);
```

This method is used to establish a value (based on strNewValue) for the cookie. There is no default value; therefore, all cookies must have their value explicitly set either via this method or by the constructor.

The setVersion( ) Method

```
public void setVersion(int nVersion);
```

This method is used to establish the version of the cookie protocol that should be used for this cookie. Version 0 is the original Netscape protocol. Version 1 is an experimental protocol. The default value is version 0.

The HttpServlet Class

```
public abstract class HttpServlet
  extends javax.servlet.GenericServlet
  Implements java.io.Serializable
{
  public HttpServlet();
  protected void doDelete(HttpServletRequest req,
    HttpServletResponse res)
```

```
    throws ServletException, IOException;
  protected void doGet(HttpServletRequest req,
    HttpServletResponse res)
    throws ServletException, IOException;
  protected void doOptions(HttpServletRequest req,
    HttpServletResponse res)
    throws ServletException, IOException;
  protected void doPost(HttpServletRequest req,
    HttpServletResponse res)
    throws ServletException, IOException;
  protected void doPut(HttpServletRequest req,
    HttpServletResponse res)
    throws ServletException, IOException;
  protected void doTrace(HttpServletRequest req,
    HttpServletResponse res)
    throws ServletException, IOException;
  protected void doGetLastModified(HttpServletRequest req,
    HttpServletResponse res)
    throws ServletException, IOException;
  public void service(ServletRequest req, ServletResponse res)
    throws ServletException, IOException;
}
```

In order to run within the context of an HTTP servlet engine, a class must extend the `HttpServlet` class. This abstract class provides methods that can be overridden by the derived class in order to enable access to particular functions. We describe these methods in the following sections.

HttpServlet( )

```
public HttpServlet();
```

The construct can be implemented within a class derived from `HttpServlet`; however, there is little reason to perform this action. Because each thread (representing a single request) will run through the same instance of the servlet, class, members are effectively a servlet resource (not a user, not a connection, and not a session resource). All servlet resources should be initialized within the overridden `GenericServlet.init()` method.

doDelete( )

```
protected void doDelete(HttpServletRequest req,
  HttpServletResponse res)
  throws ServletException, IOException;
```

The `doDelete()` method is called in sympathy to the servlet receiving an HTTP request that uses the `delete` method. Because neither WML nor WML Script provides a means of invoking the `delete` method, the `doDelete()` method is not typically implemented in a WAP Servlet. If the method is not implemented, the servlet engine will reject the requests.

doGet( )

```
protected void doGet(HttpServletRequest req,
  HttpServletResponse res)
  throws ServletException, IOException;
```

The `doGet()` method is called in sympathy to the servlet receiving an HTTP request that uses the `get` method. If the method is not implemented, the servlet engine will reject the requests.

doOptions( )

```
protected void doOptions(HttpServletRequest req,
  HttpServletResponse res)
  throws ServletException, IOException;
```

The `doOptions()` method is called in sympathy to the servlet receiving an HTTP request that uses the `options` method. Because neither WML nor WML Script provides a means of invoking the `options` method, the `doOptions()` method is not typically implemented in a WAP Servlet. If the method is not implemented, the servlet engine will reject the requests.

doPost( )

```
protected void doPost(HttpServletRequest req,
  HttpServletResponse res)
  throws ServletException, IOException;
```

The `doPost()` method is called in sympathy to the servlet receiving an HTTP request that uses the `options` method. If the method is not implemented, the servlet engine will reject the requests.

doPut( )

```
protected void doPut(HttpServletRequest req,
  HttpServletResponse res)
  throws ServletException, IOException;
```

The `doPut()` method is called in sympathy to the servlet receiving an HTTP request that uses the `put` method. Because neither WML nor WML Script provide a means of invoking the `put` method, the `doPut()` method is not typically implemented in a WAP Servlet. If the method is not implemented, the servlet engine will reject the requests.

doTrace( )

```
protected void doTrace(HttpServletRequest req,
  HttpServletResponse res)
  throws ServletException, IOException;
```

The `doTrace()` method is called in sympathy to the servlet receiving an HTTP request that uses the `trace` method. Because neither WML nor WML Script provide a means of invoking the trace method, the doTrace() method is not typically implemented in a WAP Servlet. If the method is not implemented, the servlet engine will reject the requests.

doGetLastModified( )

```
protected void doGetLastModified(HttpServletRequest req,
  HttpServletResponse res)
  throws ServletException, IOException;
```

The doGetLastModified() method is called in sympathy to the servlet receiving an HTTP request for the last date of modification on a resource.

service( )

```
public void service(ServletRequest req, ServletResponse res)
  throws ServletException, IOException;
```

The default implementation of the service() method will invoke the appropriate class method based on the HTTP request method specified. This method can be overridden if some other action is to be taken.

The HttpSessionBindingEvent Class

```
public class HttpSessionBindingEvent
  extends java.util.EventObject
{
  public HttpSessionBindingEvent(final HttpSession session,
    String strName);
  public String getName();
  public HttpSession getSession();
}
```

The HttpSessionBindingEvent( ) Constructor

```
public HttpSessionBindingEvent(final HttpSession session,
  String strName);
```

The HttpSessionBindingEvent constructor is used to instantiate a binding event to be sent to HttpSessionBindingListeners. The session argument identifies the session that binds a named resource. The strName argument identifies the name of the resource.

The getName( ) Method

```
public String getName();
```

This method can be used by an HttpSessionBindingListener in order to identify the servlet session to which a named resource has been bound.

The getSession( ) Method

```
public HttpSession getSession();
```

This method can be used by an `HttpSessionBindingListener` in order to identify the named resource that is bound to a session.

The HttpUtils Class

```
public class HttpUtils
  extends java.util.EventObject
{
  public static StringBuffer getRequestURL(HttpServletRequest
   request);
  public static java.util.Hashtable parseQueryString(String
    strQueryString);
  public static java.util.Hashtable parsePostData(int len,
    ServletInputStream in);
}
```

The `HttpUtils` class provides three general-purpose HTTP-oriented utility methods. The following sections describe those methods.

The getRequestURL( ) Method

```
public static StringBuffer getRequestURL(HttpServletRequest request);
```

The `getRequestURL()` method constructs the original URL that is associated with the HTTP request specified in the `request` argument.

The parseQueryString( ) Method

```
public static java.util.Hashtable parseQueryString(String
 strQueryString);
```

The `parseQueryString()` method is used to parse the query string specified in the `strQueryString` argument. Each name-value pair of the post content is used to populate the `Hashtable` returned. Values are stored in the hash table as strings and are accessed via their name.

The parsePostData( ) Method

```
public static java.util.Hashtable parsePostData(int len,
  ServletInputStream in);
```

The `parsePostData()` method is used to parse the content portion of a request that uses the post method. Each name-value pair of the post content is used to populate the `Hashtable` returned. Values are stored in the hash table as strings and are accessed via their name.

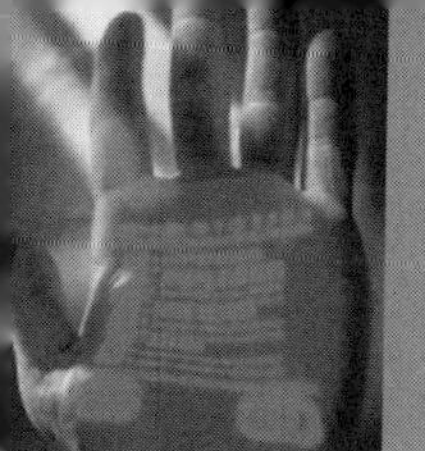

APPENDIX C

ServletEngine Sources

When developing servlets, you should have a servlet engine available for testing. There are many commercial-quality *Hypertext Transport Protocol* (HTTP) servers available (some such as the Apache are even available for free). You are certainly welcome to use any servlet engine that you desire. There are three reasons why I wrote and included the following servlet engine. First, I wanted to make sure that you had something ready to use when you opened this book. Second, I configured the servlet engine to work with the examples in the book and for working with *Wireless Application Protocol* (WAP) Servlets in general; therefore, there are no configuration hassles. Third, this application serves as an example implementation of the *Java Servlet Development Kit* (JSDK). While it is certainly possible to learn the JSDK without a look under the hood of a servlet engine, an understanding of what a server does and how it functions will certainly provide you with additional insight.

As stated in the copyright notice, I do not make any claims about the suitability of the Servlet server for a particular application; however, in general, I can say that the server was intended for nothing more than an application for driving the various examples within this book. In general, the Servlet engine hould provide you with enough functionality to develop your servlets. You are free to use the servlet engine for personal educational purposes; however, you are not licensed to use the software for profit.

The Legal Stuff

The following text appears at the top of each source file. To avoid wasting space, I do not repeatedly show this material, but it does apply to each class as follows.

```
/*******************************************************************
 *
 * Copyright (c) 1999-2000 John L. Cook, III All Rights Reserved.
 *
 * This software is confidential information ("Confidential
 * Information").  This confidential information is the property of
 * John L. Cook, III ("the Author").
 *
 * You shall not disclose such Confidential Information and shall use
 * it only in accordance with the terms of the license agreement you
 * entered with the author.
 *
 * THE AUTHOR MAKES NO REPRESENTATIONS OR WARRANTIES ABOUT THE
 * SUITABILITY OF THE SOFTWARE, EITHER EXPRESS OR IMPLIED, INCLUDING
 * BUT NOT LIMITED TO IMPLIED WARRANTIES OF MERCHANTABILITY, FITNESS
 * FOR A PARTICULAR PUPOSE, OR NON-INFRINGEMENT.  THE AUTHOR SHALL
 * NOT BE LIABLE FOR ANY DAMAGES SUFFERED BY LICENSEE AS A RESULT
 * OF USING, MODIFYING OR DISTRIBUTING THIS SOFTWARE OR ITS
 * DERIVATIVES.
 *
 * Copyright Version 1.1
 *
 *******************************************************************/
```

The Classes

All of the classes associated with the servlet engine reside in the package `com.n1guz.ServletEngine`.

The ServletEngine Class

The `ServletEngine` class is straightforward. This class simply starts a `ServletContainer` thread.

```
package com.n1guz.ServletEngine;

import java.util.*;
import java.net.*;
import java.io.*;

public class ServletEngine extends Thread
{
  public static void main(String strArgs[])
  {
    // According to the J2EE Specification, all servlets run and
    // reside within a container.  Start the container thread.

    new ServletContainer(strArgs).start();
```

```
    }
  }
```

The ServletContainer Class

The `ServletContainer` class does a bit more work. This class loads the configuration file "ServletEngine.conf," located in the /code directory of the accompanying CD-ROM in this book. The configuration file is assumed to be in the working directory. The `ServletContextImp` and `HttpSessionContextImp` classes are also created. After initialization, the container opens a `java.net.ServerSocket` and waits for connections. Once a connection is made, it is passed to a `ServletConnection` object for processing.

```
package com.n1guz.ServletEngine;

import java.util.*;
import java.net.*;
import java.io.*;

public class ServletContainer extends Thread
{
  // Default number of backlog connections.

  protected int m_nBacklog = 10;

  // Default server name.

  protected String m_strServerName = null;

  // Default server port.

  protected int m_nServerPort = 80;

  // The one and only servlet context.

  protected ServletContextImp m_servletContext =
    new ServletContextImp();

  // The one and only session context.

  protected HttpSessionContextImp m_sessionContext =
    new HttpSessionContextImp();

  public ServletContainer (String [] strArgs)
  {
  }

  /**
   * Main body of the container.
   */
```

```
public void run ()
{
  try
  {
    m_strServerName = InetAddress.getLocalHost().getHostName();
  }
  catch (IOException e)
  {
      m_servletContext.log(e, "ServletContainer:");
  }

  // Load the server's properties.

  loadProperties();

  // Bind the port.

  ServerSocket serverSocket = null;

  try
  {
    serverSocket = new ServerSocket(m_nServerPort, m_nBacklog);
  }
  catch (IOException e)
  {
      m_servletContext.log(e, "ServletContainer:");
  }

  // Wait for connections.

  if (serverSocket != null)
  {
    handleIncomingConnections(serverSocket);
  }
}

/**
 * Load server properties from the properties file.
 */

protected void loadProperties()
{
  Properties properties = new Properties();

  File file = null;
  try
  {
    file = new File("ServletEngine.conf");
    FileInputStream fis = new FileInputStream(file);

    properties.load(fis);
```

```
        fis.close();
      }
      catch (IOException e)
      {
        System.err.println("File '" + file.getAbsolutePath() +
          "': " + e.getMessage());
      }

      // Loop through the contents of the file.

      Enumeration enum = properties.propertyNames();

      String strPropertyKey = "com.n1guz.ServletEngine.";
      int nPropertyKeyLength = strPropertyKey.length();

      while (enum.hasMoreElements())
      {
        String strPropertyName = (String) enum.nextElement();
        String strPropertyValue =
          properties.getProperty(strPropertyName);

        m_servletContext.log(strPropertyName + "=" + strPropertyValue);

        if (strPropertyName.startsWith(strPropertyKey))
        {
          strPropertyName =
            strPropertyName.substring(nPropertyKeyLength);

          if (strPropertyName.equals("listen-port"))
          {
            m_nServerPort = Integer.parseInt(strPropertyValue);
          }
          else if (strPropertyName.equals("backlog"))
          {
            m_nBacklog = Integer.parseInt(strPropertyValue);
          }
          else if (strPropertyName.startsWith("mime-type."))
          {
            handleMimeProperty(strPropertyName.substring(10),
              strPropertyValue);
          }
          else if (strPropertyName.startsWith("servlet."))
          {
            handleServletProperty(strPropertyName.substring(8),
              strPropertyValue);
          }
        }
      }
    }

    /**
```

```
 * Process a mime type from the properties file.
 */

protected void handleMimeProperty(String strPropertyName,
String strPropertyValue)
{
  StringTokenizer tok = new StringTokenizer(strPropertyValue,",");

  while (tok.hasMoreTokens())
  {
    String strToken = tok.nextToken();

    m_servletContext.addMimeType(strToken, strPropertyName);
  }
}

/**
 * Process a servlet from the properties file.
 */

protected void handleServletProperty(String strPropertyName,
  String strPropertyValue)
{
  m_servletContext.addServlet(strPropertyName, strPropertyValue);
}

/**
 * Wait for requests, and kick off EngineConnection threads
 * to process them as they come in.
 */

protected void handleIncomingConnections(ServerSocket serverSocket)
{
  for (;;)
  {
    try
    {
      Socket socket = serverSocket.accept();

      EngineConnection engineConnection = new EngineConnection();

      if (engineConnection != null)
      {
          engineConnection.init(socket, m_servletContext,
              m_sessionContext, m_strServerName, m_nServerPort);

          engineConnection.start();
      }
    }
    catch (Exception e)
    {
```

```
          m_servletContext.log(e, "ServletContainer:");
        }
      }
    }
  }
```

The ServletContextImp Class

The `ServletContext` class provides an environment for a group of servlets. Because this `ServletEngine` is a single-context engine, all servlets run within the context of a single instance of this class. Servlets can use their `ServletContext` to find out about their environment and the other servlets with which they share their environment.

```
package com.n1guz.ServletEngine;

import java.util.*;
import java.io.*;
import java.net.*;
import javax.servlet.*;
import javax.servlet.http.*;

public class ServletContextImp implements ServletContext
{
  protected static String sm_strServerSoftware =
    "ServletEngine V1.0 Copyright 1999-2000 John L. Cook, III";

  protected Hashtable m_hashMimeTypes = new Hashtable();
  protected Hashtable m_hashServlets = new Hashtable();
  protected Hashtable m_hashServletObjects = new Hashtable();
  protected String m_strDocroot = ".";
  protected Hashtable m_hashAttributes = new Hashtable();

  /* *************************************************************
   **************************************************************
   * NON-EXPOSED METHODS:
   * The following methods are for internal use and not exposed
   * via the JSDK interface.
   **************************************************************
   **************************************************************/

  public void setDocroot(String strDocroot)
  {
    m_strDocroot = strDocroot;
  }

  public void addServlet(String strName, String strClass)
  {
    if (strName == null)
    {
      throw new IllegalArgumentException();
```

```
      }

      if (strClass == null)
      {
        throw new IllegalArgumentException();
      }

      m_hashServlets.put(strName, strClass);
    }

    public void addMimeType(String strExt, String strType)
    {
      if (strExt == null)
      {
        throw new IllegalArgumentException();
      }

      if (strType == null)
      {
        throw new IllegalArgumentException();
      }

      m_hashMimeTypes.put(strExt, strType);
    }

    public HttpServlet loadServlet(String strClass)
    {
      Class theClass = null;

      if (strClass != null)
      {
        try
        {
          theClass = Class.forName(strClass);
        }
        catch (Exception e)
        {
          System.err.println("Unable to load servlet " + strClass);
          theClass = null;
        }
      }

      // Get the servlet class.

      HttpServlet httpServlet = null;

      if (theClass != null)
      {
        try
        {
          httpServlet = (HttpServlet) theClass.newInstance();
```

```
      }
      catch (InstantiationException e)
      {
        httpServlet = null;
      }
      catch (IllegalAccessException e)
      {
        httpServlet = null;
      }
    }

    if (httpServlet != null)
    {
      try
      {
        ServletConfig servletConfig = new ServletConfigImp(this);

        httpServlet.init(servletConfig);
      }
      catch (Exception e)
      {
      }
    }

    return httpServlet;
  }

  public HttpServlet findServlet(String strPath,
  HttpServletRequestImp request)
  {
    // Assume that the path refers to a servlet

    Enumeration enum = m_hashServlets.keys();

    boolean bFound = false;
    String strKey = null;

    while (enum.hasMoreElements())
    {
      strKey = (String) enum.nextElement();

      if (strPath.startsWith(strKey))
      {
        int nLenPath = strPath.length();
        int nLenKey = strKey.length();

        if (nLenPath != nLenKey)
        {
          String strPathInfo = strPath.substring(nLenKey);
          if (strPathInfo.startsWith("/"))
          {
```

```
          request.setPathInfo(strPathInfo);
          request.setServletPath(strKey);
          bFound = true;
          break;
        }
      }
      else
      {
        request.setPathInfo("");
        request.setServletPath(strKey);
        bFound = true;
        break;
      }
    }
  }

  String strClass = null;

  // If a servlet class was not found, assume that a document
  // was requested, and use the MimeServlet.

  if (bFound)
  {
    strClass = (String) m_hashServlets.get(strKey);
  }
  else
  {
    strClass = "com.n1guz.ServletEngine.MimeServlet";
  }

  // See if an instance of the servlet already exists.

  HttpServlet servlet =
    (HttpServlet) m_hashServletObjects.get(strClass);

  if (servlet == null)
  {
    // No so create an instance.

    servlet = loadServlet(strClass);

    if (servlet != null)
    {
      m_hashServletObjects.put(strClass, servlet);
    }
  }

  return servlet;
}

/* ******************************************************************
```

```
 ****************************************************************
 * EXPOSED METHODS:
 * The following methods are required and exposed by the JSDK
 * interface.
 ****************************************************************
 ****************************************************************/

/**
 * Returns the value of the named attribute of the network
 * service, or null if the attribute does not exist.  This
 * method allows access to additional information about the
 * service, not already provided by the other methods in
 * this interface. Attribute names should follow the same
 * convention as package names.  The package names java.* and
 * javax.* are reserved for use by Javasoft, and
 * com.sun.* is reserved for use by Sun Microsystems.
 *
 * @param name the name of the attribute whose value is required
 * @return the value of the attribute, or null if the attribute
 * does not exist.
 */

public Object getAttribute(String strAttribute)
{
  return m_hashAttributes.get(strAttribute);
}

/**
 * Returns an <code>Enumeration</code> containing the
 * attribute names available
 * within this servlet context. You can use the
 * {@link #getAttribute} method with an attribute name
 * to get the value of an attribute.
 *
 * @return           an <code>Enumeration</code> of attribute
 *                names
 *
 * @see           #getAttribute
 *
 */

public Enumeration getAttributeNames()
{
  return m_hashAttributes.keys();
}

/**
 * Returns a <code>ServletContext</code> object that
 * corresponds to a specified URL on the server.
 *
 * <p>This method allows servlets to gain
```

```
 * access to the resources located at a specified URL and obtain
 * {@link RequestDispatcher} objects from it.
 *
 * <p>In security conscious environments, the servlet engine can
 * return <code>null</code> for a given URL.
 *
 * @param uripath      a <code>String</code> specifying the URL for
 *              which you are requesting a
 *                  <code>ServletContext</code> object
 *
 * @return          the <code>ServletContext</code> object that
 *              corresponds to the named URL
 *
 * @see            RequestDispatcher
 *
 */

public ServletContext getContext(String strArg)
{
  return this;
}

/**
 * Returns the major version of the Java Servlet API that this
 * Web server supports. All implementations that comply
 * with Version 2.1 must have this method
 * return the integer 2.
 *
 * @return           2
 *
 */

public int getMajorVersion()
{
  return 2;
}

/**
 * Returns the mime type of the specified file, or null if not
 * known.
 *
 * @param file name of the file whose mime type is required
 */

public String getMimeType(String strExt)
{
  if (strExt == null)
  {
    throw new IllegalArgumentException();
  }
```

```
  return (String) m_hashMimeTypes.get(strExt);
}

/**
 * Returns the minor version of the Servlet API that this
 * Web server supports. All implementations that comply
 * with Version 2.1 must have this method
 * return the integer 1.
 *
 * @return          1
 *
 */

public int getMinorVersion()
{
  return 1;
}

/**
 * Applies alias rules to the specified virtual path and returns
 * the corresponding real path.  For example, in an HTTP servlet,
 * this method would resolve the path against the HTTP service's
 * docroot.  Returns null if virtual paths are not supported, or
 * if the translation could not be performed for any reason.
 *
 * @param path the virtual path to be translated into a real path
 */

public String getRealPath(String strPath)
{
  File file = new File(m_strDocroot, strPath);

  return file.getAbsolutePath();
}

/**
 *
 * Returns a {@link RequestDispatcher} object that acts
 * as a wrapper for the resource located at the named path.
 * You can use a <code>RequestDispatcher</code> object to forward
 * a request to the resource or include a resource in a response.
 *
 * <p>The pathname must be in the form
 * <code>/dir/dir/file.ext</code>.
 * This method returns <code>null</code> if the
 * <code>ServletContext</code>
 * cannot return a <code>RequestDispatcher</code>.
 *
 * <p>The servlet engine is responsible for wrapping the resource
 * with a <code>RequestDispatcher</code> object.
```

```
 *
 * @param urlpath       a <code>String</code> specifying the pathname
 *                      to the resource
 *
 * @return              a <code>RequestDispatcher</code> object
 *                      that acts as a wrapper for the resource
 *                      at the path you specify
 *
 * @see                 RequestDispatcher
 *
 */

public RequestDispatcher getRequestDispatcher(String strArg)
{
  return null;
}

/**
 * Returns the resource that is mapped to a specified
 * path. The path must be in the form
 * <code>/dir/dir/file.ext</code>.
 *
 * <p>This method allows the Web
 * server to make a resource available to a servlet from
 * any source. Resources
 * can be located on a local or remote
 * file system, in a database, or on a remote network site.
 *
 * <p>This method can return <code>null</code>
 * if no resource is mapped to the pathname.
 *
 * <p>The servlet engine must implement the URL handlers
 * and <code>URLConnection</code> objects that are necessary
 * to access the resource.
 *
 * <p>This method has a different purpose than
 * <code>java.lang.Class.getResource</code>,
 * which looks up resources based on a class loader. This
 * method does not use class loaders.
 *
 * @param path      a <code>String</code> specifying
 *               the path to the resource,
 *                in the form <code>/dir/dir/file.ext</code>
 *
 * @return             the resource located at the named path,
 *                or <code>null</code> if there is no resource
 *               at that path
 *
 * @exception MalformedURLException
 *                   if the pathname is not given in
 *                the correct form
```

```
 *
 */

public URL getResource(String strResource)
{
  return null;
}

/**
 * Returns the resource located at the named path as
 * an <code>InputStream</code> object.
 *
 * <p>The data in the <code>InputStream</code> can be
 * of any type or length. The path must be of
 * the form <code>/dir/dir/file.ext</code>. This method
 * returns <code>null</code> if no resource exists at
 * the specified path.
 *
 * <p>Metainformation such as content length and content type
 * that is available when you use the <code>getResource</code>
 * method is lost when you use this method.
 *
 * <p>The servlet engine must implement the URL handlers
 * and <code>URLConnection</code> objects necessary to access
 * the resource.
 *
 * <p>This method is different from
 * <code>java.lang.Class.getResourceAsStream</code>,
 * which uses a class loader. This method allows servlet engines
 * to make a resource available
 * to a servlet from any location, without using a class loader.
 *
 *
 * @param name      a <code>String</code> specifying the path
 *               to the resource,
 *                in the form <code>/dir/dir/file.ext</code>
 *
 * @return           the <code>InputStream</code> returned to the
 *               servlet, or <code>null</code> if no resource
 *               exists at the specified path
 *
 *
 */

public InputStream getResourceAsStream(String strResource)
{
  return null;
}

/**
 * Returns the name and version of the network service under
```

```
 * which the servlet is running. For example, if the network
 * service was an HTTP service, then this would be the same
 * as the CGI variable SERVER_SOFTWARE.
 */

public String getServerInfo()
{
  return sm_strServerSoftware;
}

/**
 * Returns the servlet of the specified name, or null if not
 * found.  When the servlet is returned it is initialized and
 * ready to accept service requests.
 *
 * Note: This is a dangerous method to call for the
 * following reasons:
 *
 * 1. When this method is called the state of the servlet
 *    may not be known, and this could cause problems with
 *    the server's servlet state machine.
 *
 * 2. It is a security risk to allow any servlet to be able
 *    to access the methods of another servlet.
 *
 * @param name the name of the desired servlet
 * @exception ServletException if the servlet could not be
 * initialized.
 */

public Servlet getServlet(String strServlet)
{
  if (strServlet == null)
  {
    throw new IllegalArgumentException();
  }

  return (Servlet) m_hashServlets.get(strServlet);
}

/**
 * Returns an enumeration of the Servlet object names in this
 * server.  Only servlets that are accessible (i.e., from the
 * same namespace) will be returned.  The enumeration always
 * includes the servlet itself.
 *
 * 1. When this method is called the state of the servlet
 *    may not be known, and this could cause problems with
 *    the server's servlet state machine.
 *
 * 2. It is a security risk to allow any servlet to be able
```

```
 *    to access the methods of another servlet.
 */

public Enumeration getServletNames()
{
  return m_hashServlets.keys();
}

/**
 * Returns an enumeration of the Servlet objects in this
 * server.  Only servlets that are accessible (i.e., from
 * the same namespace) will be returned.  The enumeration
 * always includes the servlet itself.
 *
 * 1. When this method is called the state of the servlet
 *    may not be known, and this could cause problems with
 *    the server's servlet state machine.
 *
 * 2. It is a security risk to allow any servlet to be able
 *    to access the methods of another servlet.
 *
 * @deprecated
 * Please use getServletNames in conjunction with getServlet
 * @see #getServletNames
 * @see #getServlet
 */

public Enumeration getServlets()
{
  return m_hashServlets.elements();
}

/**
 * Writes the given message string to the servlet log file.
 * The name of the servlet log file is server specific; it
 * is normally an event log.
 * @param msg the message to be written
 */

public void log(String strMsg)
{
  System.out.println(strMsg);
}

/**
 * Write the stacktrace and the given message string to the
 * servlet log file. The name of the servlet log file is
 * server specific; it is normally an event log.
 * @param exception the exception to be written
 * @param msg the message to be written
 */
```

```
public void log(Exception e, String strMsg)
{
  e.printStackTrace(System.out);
  log(strMsg);
}

public void log(String strMsg, Throwable throwable)
{
  System.out.println(strMsg);
  throwable.printStackTrace(System.out);
}

/**
 * Removes the attribute with the given name from
 * the servlet context. If you remove an attribute, and
 * then use {@link #getAttribute} to retrieve the
 * attribute's value, <code>getAttribute</code> returns
 * <code>null</code>.
 *
 *
 * @param name     a <code>String</code> specifying the name
 *                 of the attribute to be removed
 *
 */

public void removeAttribute(String strName)
{
  m_hashAttributes.remove(strName);
}

/**
 *
 * Gives an attribute a name in this servlet context. If
 * the name specified is already used for an attribute, this
 * method will overwrite the old attribute and bind the name
 * to the new attribute.
 *
 * <p>Attribute names should follow the same convention as package
 * names. The Java Servlet API specification reserves names
 * matching <code>java.*</code>, <code>javax.*</code>, and
 * <code>sun.*</code>.
 *
 *
 * @param name      a <code>String</code> specifying the name
 *                of the attribute
 *
 *
 * @param object      an <code>Object</code> representing the
 *                attribute to be given the name
 *
 *
```

```
    *
    */

   public void setAttribute(String strName, Object obj)
   {
     m_hashAttributes.put(strName, obj);
   }
}
```

The HttpSessionContextImp Class

The `HttpSessionContextImp` class maintains the list of HTTP sessions in play.

```
package com.n1guz.ServletEngine;

import javax.servlet.*;
import javax.servlet.http.*;
import java.util.*;
import java.net.*;
import java.io.*;

public class HttpSessionContextImp extends Hashtable
implements HttpSessionContext
{
  public Enumeration getIds()
  {
    return keys();
  }

  public HttpSession getSession(String strId)
  {
    return (HttpSession) get(strId);
  }
}
```

The ServerConnection Class

The `ServerConnection` class is used to process a request. This class initializes an `HttpServletRequestImp` and `HttpServletResponseImp` object and calls the service method of the servlet. If the servlet creates a new `HttpSession` object during the request, the session is stored for future use.

```
package com.n1guz.ServletEngine;

import javax.servlet.*;
import javax.servlet.http.*;
import java.util.*;
import java.net.*;
import java.io.*;
```

```
public class EngineConnection extends Thread
{
  protected Socket m_socket = null;
  protected static Hashtable sm_hashSessions = new Hashtable();
  protected static Hashtable sm_hashServlets = null;
  protected ServletContextImp m_servletContext = null;
  protected HttpSessionContextImp m_sessionContext = null;
  protected String m_strServerName = null;
  protected int m_nServerPort = 0;

  public EngineConnection()
  {
  }

  public EngineConnection(Socket socket,
  ServletContextImp servletContext,
  HttpSessionContextImp sessionContext,
  String strServerName,int nServerPort)
  {
    init(socket, servletContext, sessionContext,
    strServerName, nServerPort);
  }

  public void init(Socket socket, ServletContextImp servletContext,
  HttpSessionContextImp sessionContext,
  String strServerName,int nServerPort)
  {
    m_socket = socket;
    m_servletContext = servletContext;
    m_sessionContext = sessionContext;
    m_strServerName = strServerName;
    m_nServerPort = nServerPort;
  }

  public void run()
  {
    try
    {
      // Create and initialize an HTTPRequest object.

      HttpServletRequestImp httpRequest =
      new HttpServletRequestImp(m_sessionContext);

      httpRequest.setServerName(m_strServerName);

      httpRequest.setServerPort(m_nServerPort);
      httpRequest.loadFrom(m_socket);

      InetAddress inetAddress = m_socket.getInetAddress();
      httpRequest.setRemoteAddr(inetAddress.getHostAddress());
```

```
// Get the path info.

String strPath = httpRequest.getPathInfo();

// Get the servlet class.

HttpServlet httpServlet =
m_servletContext.findServlet(strPath, httpRequest);

if (httpServlet != null)
{
  HttpServletResponseImp httpResponse =
  new HttpServletResponseImp(httpRequest);

  httpResponse.setStatus(HttpServletResponse.SC_OK);

  HttpSessionImp session = null;

  Cookie cookies[] = httpRequest.getCookies();

  int nCount = cookies.length;
  for (int nIndex = 0; nIndex < nCount; nIndex++)
  {
    if (cookies[nIndex].getName().equals("ServletSessionID"))
    {
      String strSessionID = cookies[nIndex].getValue();
      session = (HttpSessionImp)
        sm_hashSessions.get(strSessionID);

      if (session != null)
      {
        httpRequest.setSessionFromCookie(session);
      }
      break;
    }
  }

  if (session == null)
  {
    String strSessionID =
      httpRequest.getParameter("ServletSessionID");

    if (strSessionID != null)
    {
      session = (HttpSessionImp)
        sm_hashSessions.get(strSessionID);

      if (session != null)
      {
        httpRequest.
```

```
              setSessionFromParameter(session);
            }
          }
        }

        httpServlet.service(httpRequest,httpResponse);

        session = (HttpSessionImp)
        httpRequest.getSession(false);

        if ((session != null) && session.isValid())
        {
          if (session.isNew())
          {
            String strSessionID = session.getId();
            Cookie cookie =
              new Cookie("ServletSessionID", strSessionID);
            cookie.setPath("/");
            cookie.setMaxAge(60*60*12);
            httpResponse.addCookie(cookie);

            sm_hashSessions.put(strSessionID,session);
          }
          if (session instanceof HttpSessionImp)
          {
            ((HttpSessionImp)session).touch();
          }
        }

        httpResponse.send(m_socket);
      }

      m_socket.close();
    }
    catch (IOException e)
    {
      synchronized(System.out)
      {
        System.out.println("Session error: " + e);
      }
    }
    catch (ServletException e)
    {
      synchronized(System.out)
      {
        System.out.println("Session error: " + e);
      }
    }
  }
}
```

The HttpSessionImp Class

The `HttpSessionImp` class provides the servlet with a notion of a session. The servlet can use this class for saving session-persistent information between requests.

```
package com.n1guz.ServletEngine;

import javax.servlet.*;
import javax.servlet.http.*;
import java.util.*;
import java.net.*;
import java.io.*;

public class HttpSessionImp implements HttpSession
{
  protected Date m_dateCreated = new Date();
  protected Date m_dateLastAccessed = m_dateCreated;
  protected Hashtable m_hashValues = new Hashtable();
  protected boolean m_bIsNew = true;
  protected String m_strID = null;
  protected boolean m_bValid = false;
  protected HttpSessionContextImp m_sessionContext = null;
  protected int m_nMaxInactiveInterval = 0;

  /* ***************************************************************
   ***************************************************************
   * NON-EXPOSED METHODS:
   * The following methods are for internal use and not exposed
   * via the JSDK interface.
   ***************************************************************
   ***************************************************************/

  public HttpSessionImp(HttpSessionContextImp sessionContext)
  {
    m_strID =
      Long.toHexString(Double.doubleToLongBits(Math.random())) +
      Long.toHexString(m_dateCreated.getTime());

    m_bValid = true;
    m_sessionContext = sessionContext;
  }

  public void touch()
  {
    m_bIsNew = false;
    m_dateLastAccessed = new Date();
  }

  public boolean isValid()
  {
```

```
    return m_bValid;
  }

  /* ***************************************************************
   ***************************************************************
   * EXPOSED METHODS:
   * The following methods are required and exposed by the JSDK
   * interface.
   ***************************************************************
   ***************************************************************/

  /**
   * Returns the time at which this session representation was
   * created, in milliseconds since midnight, January 1, 1970 UTC.
   *
   * @return the time when the session was created
   * @exception IllegalStateException if an attempt is made to access
   * session data after the session has been invalidated
   */

  public long getCreationTime ()
  {
    if (!m_bValid)
    {
      throw new IllegalStateException("HttpSession.getValue(): " +
        "invalid session");
    }

    return m_dateCreated.getTime();
  }

  /**
   * Returns the identifier assigned to this session. An
   * HttpSession's identifier is a unique string that is
   * created and maintained by HttpSessionContext.
   *
   * @return the identifier assigned to this session
   * @exception IllegalStateException if an attempt is made to access
   * session data after the session has been invalidated
   */

  public String getId ()
  {
    if (!m_bValid)
    {
      throw new IllegalStateException("HttpSession.getValue(): " +
        "invalid session");
    }

    return m_strID;
  }
```

```
/**
 * Returns the last time the client sent a request carrying the
 * identifier assigned to the session. Time is expressed
 * as milliseconds since midnight, January 1, 1970 UTC.
 * Application level operations, such as getting or setting a value
 * associated with the session, does not affect the access time.
 *
 * <P> This information is particularly useful in session
 * management policies.  For example,
 * <UL>
 * <LI>a session manager could leave all sessions
 * which have not been used in a long time
 * in a given context.
 * <LI>the sessions can be sorted according to age to optimize
 * some task.
 * </UL>
 *
 * @return the last time the client sent a request carrying
 * the identifier assigned to the session
 * @exception IllegalStateException if an attempt is made to
 * access session data after the session has been invalidated
 */

public long getLastAccessedTime ()
{
  if (!m_bValid)
  {
    throw new IllegalStateException("HttpSession.getValue(): " +
      "invalid session");
  }

  return m_dateLastAccessed.getTime();
}

public int getMaxInactiveInterval()
{
  return m_nMaxInactiveInterval;
}

/**
 * Returns the context in which this session is bound.
 *
 * @return the name of the context in which this session is bound
 * @exception IllegalStateException if an attempt is made to access
 * session data after the session has been invalidated
 */

public HttpSessionContext getSessionContext ()
{
  if (!m_bValid)
  {
```

```
      throw new IllegalStateException("HttpSession.getValue(): " +
        "invalid session");
    }

    return m_sessionContext;
  }

  /**
   * Returns the object bound to the given name in the session's
   * application layer data.  Returns null if there is no such
   * binding.
   *
   * @param name the name of the binding to find
   * @return the value bound to that name, or null if the binding
   * does not exist.
   * @exception IllegalStateException if an attempt is made to access
   * HttpSession's session data after it has been invalidated
   */

  public Object getValue (String strName)
  {
    if (!m_bValid)
    {
      throw new IllegalStateException("HttpSession.getValue(): " +
        "invalid session");
    }
    return m_hashValues.get(strName);
  }

  /**
   * Returns an array of the names of all the application layer
   * data objects bound into the session. For example, if you want
   * to delete all of the data objects bound into the session, use
   * this method to obtain their names.
   *
   * @return an array containing the names of all of the application
   * layer data objects bound into the session
   * @exception IllegalStateException if an attempt is made to access
   * session data after the session has been invalidated
   */

  public String [] getValueNames ()
  {
    if (!m_bValid)
    {
      throw new IllegalStateException("HttpSession.getValue(): " +
        "invalid session");
    }

    String strRetval[] = null;
```

```
    synchronized (m_hashValues)
    {
      int nSize = m_hashValues.size();

      strRetval = new String[nSize];

      Enumeration enum = m_hashValues.keys();

      int nCount = 0;

      while (enum.hasMoreElements())
      {
        String strKey = (String) enum.nextElement();

        strRetval[nCount++] = strKey;
      }
    }

    return strRetval;
  }

  /**
   * Causes this representation of the session to be invalidated and
   * removed from its context.
   *
   * @exception IllegalStateException if an attempt is made to access
   * session data after the session has been invalidated
   */

  public void invalidate ()
  {
    if (!m_bValid)
    {
      throw new IllegalStateException("HttpSession.getValue(): " +
        "invalid session");
    }

    m_bValid = false;
  }

  /**
   * A session is considered to be "new" if it has been created
   * by the server, but the client has not yet acknowledged
   * joining the session. For example, if the server supported
   * only cookie-based sessions and the client had completely
   * disabled the use of cookies, then calls to
   * HttpServletRequest.getSession() would always return "new"
   * sessions.
   *
   * @return true if the session has been created by the server but
```

```
 * the client has not yet acknowledged joining the session; false
 * otherwise
 * @exception IllegalStateException if an attempt is made to access
 * session data after the session has been invalidated
 */

public boolean isNew ()
{
  if (!m_bValid)
  {
    throw new IllegalStateException("HttpSession.getValue(): " +
      "invalid session");
  }

  return m_bIsNew;
}

/**
 * Binds the specified object into the session's application layer
 * data with the given name.  Any existing binding with the same
 * name is replaced.  New (or existing) values that implement the
 * HttpSessionBindingListener interface will call its
 * valueBound() method.
 *
 * @param name the name to which the data object will be bound.
 * This parameter cannot be null.
 * @param value the data object to be bound.  This parameter
 * cannot be null.
 * @exception IllegalStateException if an attempt is made to access
 * session data after the session has been invalidated
 */

public void putValue (String strName, Object objValue)
{
  if (!m_bValid)
  {
    throw new IllegalStateException("HttpSession.getValue(): " +
      "invalid session");
  }

  m_hashValues.put(strName, objValue);
}

/**
 * Removes the object bound to the given name in the session's
 * application layer data.  Does nothing if there is no object
 * bound to the given name.  The value that implements the
 * HttpSessionBindingListener interface will call its
 * valueUnbound() method.
 *
 * @param name the name of the object to remove
```

```
     * @exception IllegalStateException if an attempt is made to access
     * session data after the session has been invalidated
     */

    public void removeValue (String strName)
    {
      if (!m_bValid)
      {
        throw new IllegalStateException("HttpSession.getValue(): " +
          "invalid session");
      }

      m_hashValues.remove(strName);
    }

    public void setMaxInactiveInterval(int nMaxInactiveInterval)
    {
      m_nMaxInactiveInterval = nMaxInactiveInterval;
    }
  }
```

The ServletConfigImp Class

The ServletConfigImp class provides the servlet with a means of obtaining its configuration information (initialization parameters). The class also provides internal methods for establishing these values.

```
package com.n1guz.ServletEngine;

import java.util.*;
import javax.servlet.*;
import javax.servlet.http.*;

public class ServletConfigImp implements ServletConfig
{
  protected ServletContext m_servletContext = null;
  protected Hashtable m_hashInitParameters = new Hashtable();

  /* *****************************************************************
   *****************************************************************
   * NON-EXPOSED METHODS:
   * The following methods are for internal use and not exposed
   * via the JSDK interface.
   *****************************************************************
   *****************************************************************/

  public ServletConfigImp(ServletContext servletContext)
  {
    m_servletContext = servletContext;
  }
```

```
public void addInitParameter(String strName, String strValue)
{
  m_hashInitParameters.put(strName, strValue);
}

/* ****************************************************************
*****************************************************************
 * EXPOSED METHODS:
 * The following methods are required and exposed by the JSDK
 * interface.
*****************************************************************
****************************************************************/

/**
 *
 * Returns a string containing the value of the named
 * initialization parameter of the servlet, or null if the
 * parameter does not exist.  Init parameters have a single string
 * value; it is the responsibility of the servlet writer to
 * interpret the string.
 *
 * @param name the name of the parameter whose value is requested
 */

public String getInitParameter(String strParameter)
{
  String strRetval = null;

  if (strParameter == null)
  {
    throw new IllegalArgumentException();
  }

  try
  {
    strRetval =
      (String) m_hashInitParameters.get(strParameter);
  }
  catch (Exception e)
  {
    strRetval = null;
  }

  return strRetval;
}

/**
 * Returns the names of the servlet's initialization parameters
 * as an enumeration of strings, or an empty enumeration if there
 * are no initialization parameters.
 */
```

```
    public Enumeration getInitParameterNames()
    {
      return m_hashInitParameters.keys();
    }

    /**
     * Returns the context for the servlet.
     */

    public ServletContext getServletContext()
    {
      return m_servletContext;
    }
  }
```

The ServletRequestImp Class

The `ServletRequestImp` class is used as the object that implements the `ServletRequest` interface. The `HttpServletRequestImp` class extends this class. An instance of the `HttpServletRequestImp` class is passed to the `HttpServlet.service()` method on a per-request basis.

```
package com.n1guz.ServletEngine;

import javax.servlet.*;
import javax.servlet.http.*;
import java.util.*;
import java.net.*;
import java.io.*;

public class ServletRequestImp implements ServletRequest
{
  /**
   * The length of the request content.
   */

  protected int m_nContentLength = -1;

  /**
   * The MIME type of the content.
   */

  protected String m_strContentType = null;

  /**
   * The protocol and version.
   */

  protected String m_strProtocol = null;
```

```
/**
 * The scheme (such as http or https).
 */

protected String m_strScheme = "http";

/**
 * The name of the server.
 */

protected String m_strServerName = null;

/**
 * The server port.
 */

protected int m_nServerPort = 0;

/**
 * The remote client's address.
 */

protected String m_strRemoteAddr = null;

/**
 * The DNS of the remote client.
 */

protected String m_strRemoteHost = null;

/**
 * Hashtable of parameters.
 */

protected Hashtable m_hashParameters = null;

/**
 * Content input stream.
 */

protected InputStream m_is = null;

/**
 * Input Stream Reader wrapper for the content input stream.
 */

protected InputStreamReader m_isr = null;

/**
 * Buffered Reader wrapper for the content input stream.
 */
```

```
protected BufferedReader m_br = null;

/**
 * Buffered input stream wrapper for the input stream.
 */

protected BufferedInputStream m_bis = null;

/**
 * Hashtable for storing request attributes.
 */

protected Hashtable m_attributes = new Hashtable();

/* **************************************************************
 ***************************************************************
 * NON-EXPOSED METHODS:
 * The following methods are for internal use and not exposed
 * via the JSDK interface.
 ***************************************************************
 **************************************************************/

public ServletRequestImp()
{
  m_hashParameters = new Hashtable();
}

public void setServerName (String strServerName)
{
  if (strServerName == null)
  {
    throw new IllegalArgumentException(
    "HTTPRequest.strServerName(): null server name");
  }

  m_strServerName = strServerName;
}

public void setServerPort (int nServerPort)
{
  m_nServerPort = nServerPort;
}

public void setRemoteHost(String strRemoteHost)
{
  m_strRemoteHost = strRemoteHost;
}

public void setRemoteAddr(String strRemoteAddr)
{
  m_strRemoteAddr = strRemoteAddr;
```

```
}

/* *************************************************************
 **************************************************************
 * EXPOSED METHODS:
 * The following methods are required and exposed by the JSDK
 * interface.
 **************************************************************
 *************************************************************/

/**
 * Returns the value of the named attribute of the request, or
 * null if the attribute does not exist.  This method allows
 * access to request information not already provided by the other
 * methods in this interface.  Attribute names should follow the
 * same convention as package names.
 *
 * @param name the name of the attribute whose value is required
 */

public Object getAttribute(String name)
{
  return m_attributes.get(name);
}

/**
 * Returns an <code>Enumeration</code> containing the
 * names of the attributes available to this request.
 * This method returns an empty <code>Enumeration</code>
 * if the request has no attributes available to it.
 *
 *
 * @return   an <code>Enumeration</code> of strings
 *           containing the names
 *           of the request's attributes
 */

public Enumeration getAttributeNames()
{
  return m_attributes.keys();
}

/**
 * Returns the name of the character encoding style used in this
 * request. This method returns <code>null</code> if the request
 * does not use character encoding.
 *
 *
 * @return   a <code>String</code> containing the name of
 *           the chararacter encoding style, or <code>null</code>
 *           if the request does not use character encoding
```

```
 */

public String getCharacterEncoding ()
{
  String strEncoding = null;

  if (m_isr != null)
  {
    strEncoding = m_isr.getEncoding();
  }

  return strEncoding;
}

/**
 * Returns the length, in bytes, of the content contained in the
 * request and sent by way of the input stream or -1 if the
 * length is not known. Same as the value
 * of the CGI variable CONTENT_LENGTH.
 *
 * @return    an integer containing the length of the content
 *            in the request or -1 if the length is not known
 */

public int getContentLength()
{
  return m_nContentLength;
}

/**
 * Returns the MIME type of the content of the request, or
 * <code>null</code> if the type is not known. Same as the value
 * of the CGI variable CONTENT_TYPE.
 *
 * @return    a <code>String</code> containing the name
 *            of the MIME type of
 *            the request, or -1 if the type is not known
 */

public String getContentType()
{
  return m_strContentType;
}

/**
 * Returns an input stream for reading binary data in the request
 * body.
 *
 * @see getReader
 * @exception IllegalStateException if getReader has been
 *     called on this same request.
```

```
 * @exception IOException on other I/O related errors.
 */

public ServletInputStream getInputStream() throws IOException
{
  return null;
}

/**
 * Returns a string containing the lone value of the specified
 * parameter, or null if the parameter does not exist. For example,
 * in an HTTP servlet this method would return the value of the
 * specified query string parameter. Servlet writers should use
 * this method only when they are sure that there is only one value
 * for the parameter.  If the parameter has (or could have)
 * multiple values, servlet writers should use
 * getParameterValues. If a multiple valued parameter name is
 * passed as an argument, the return value is implementation
 * dependent.
 *
 * @see #getParameterValues
 *
 * @param name the name of the parameter whose value is required.
 */

public String getParameter(String name)
{
  String strRetval = null;
  Object objParameterValue = m_hashParameters.get(name);

  if (objParameterValue != null)
  {
    if (objParameterValue instanceof String)
    {
      strRetval = (String) objParameterValue;
    }
    else if (objParameterValue instanceof Vector)
    {
      Vector vParameterValues = (Vector) objParameterValue;

      if (vParameterValues.size() > 0)
      {
        strRetval = (String) vParameterValues.elementAt(0);
      }
    }
  }

  return strRetval;
}

/**
```

```
 * Returns the parameter names for this request as an enumeration
 * of strings, or an empty enumeration if there are no parameters
 * or the input stream is empty.  The input stream would be empty
 * if all the data had been read from the stream returned by the
 * method getInputStream.
 */

public Enumeration getParameterNames()
{
  return m_hashParameters.keys();
}

/**
 * Returns the values of the specified parameter for the request as
 * an array of strings, or null if the named parameter does not
 * exist. For example, in an HTTP servlet this method would return
 * the values of the specified query string or posted form as an
 * array of strings.
 *
 * @param name the name of the parameter whose value is required.
 * @see javax.servlet.ServletRequest#getParameter
 */

public String[] getParameterValues(String name)
{
  String strRetvals[] = null;

  Object objParameterValue = m_hashParameters.get(name);

  if (objParameterValue != null)
  {
    if (objParameterValue instanceof String)
    {
      strRetvals = new String[1];
      strRetvals[0] = (String) objParameterValue;
    }
    else if (objParameterValue instanceof Vector)
    {
      Vector vParameterValues = (Vector) objParameterValue;

      int nCount = vParameterValues.size();

      strRetvals = new String[nCount];

      for (int nIndex = 0; nIndex < nCount; nIndex++)
      {
        strRetvals[nIndex] =
        (String) vParameterValues.elementAt(nIndex);
      }
    }
  }
```

```
    return strRetvals;
  }

  /**
   * Returns the name and version of the protocol the request uses
   * in the form <i>protocol/majorVersion.minorVersion</i>, for
   * example, HTTP/1.1. The value
   * returned is the same as the value of the CGI variable
   * <code>SERVER_PROTOCOL</code>.
   *
   * @return     a <code>String</code> containing the protocol
   *             name and version number
   */

  public String getProtocol()
  {
    return m_strProtocol;
  }

  /**
   * Returns a buffered reader for reading text in the request body.
   * This translates character set encodings as appropriate.
   *
   * @see getInputStream
   *
   * @exception UnsupportedEncodingException if the character set
   *  encoding is unsupported, so the text can't be correctly
   *  decoded.
   * @exception IllegalStateException if getInputStream has been
   *     called on this same request.
   * @exception IOException on other I/O related errors.
   */

  public BufferedReader getReader () throws IOException
  {
    if (m_bis != null)
    {
      throw new IllegalStateException(
      "previous call to getInputStream");
    }

    m_isr = new InputStreamReader(m_is);
    m_br = new BufferedReader(m_isr);

    return m_br;
  }

  /**
   * Applies alias rules to the specified virtual path and returns
   * the corresponding real path, or null if the translation can not
   * be performed for any reason.  For example, an HTTP servlet would
```

```
 * resolve the path using the virtual docroot, if virtual hosting
 * is enabled, and with the default docroot otherwise.  Calling
 * this method with the string "/" as an argument returns the
 * document root.
 *
 * @param path the virtual path to be translated to a real path
 */

public String getRealPath(String path)
{
  return null;
}

/**
 * Returns the Internet Protocol (IP) address of the client
 * that sent the request.
 * Same as the value of the CGI variable <code>REMOTE_ADDR</code>.
 *
 * @return      a <code>String</code> containing the
 *              IP address of the client that sent the request
 *
 */

public String getRemoteAddr()
{
  return m_strRemoteAddr;
}

/**
 * Returns the fully qualified name of the client that sent the
 * request. Same as the value of the CGI variable
 * <code>REMOTE_HOST</code>.
 *
 * @return      a <code>String</code> containing the fully
 *              qualified name of the client
 *
 */

public String getRemoteHost()
{
  return m_strRemoteHost;
}

/**
 * Returns the scheme of the URL used in this request, for example
 * "http", "https", or "ftp".  Different schemes have different
 * rules for constructing URLs, as noted in RFC 1738.  The URL used
 * to create a request may be reconstructed using this scheme, the
 * server name and port, and additional information such as URIs.
 */
```

```
public String getScheme()
{
  return m_strScheme;
}

/**
 * Returns the host name of the server that received the request.
 * Same as the CGI variable SERVER_NAME.
 */

public String getServerName()
{
  return m_strServerName;
}

/**
 * Returns the port number on which this request was received.
 * Same as the CGI variable SERVER_PORT.
 */

public int getServerPort()
{
  return m_nServerPort;
}

/**
 *
 * Stores an attribute in the context of this request.
 * Attributes are reset between requests.
 *
 * <p>Attribute names should follow the same conventions as
 * package names. Names beginning with <code>java.*</code>,
 * <code>javax.*</code>, and <code>com.sun.*</code>, are
 * reserved for use by Sun Microsystems.
 *
 *
 * @param key             a <code>String</code> specifying
 *                      the name of the attribute
 *
 * @param o             an <code>Object</code> containing
 *                      the context of the request
 *
 * @exception IllegalStateException
 *                      if the specified attribute already has
 *                      a value
 *
 */

public void setAttribute(String strName, Object obj)
{
  m_attributes.put(strName,obj);
```

```
    }
  }
```

The HttpServletRequestImp Class

The `HttpServletRequestImp` class is used as the object that implements the `HttpServletRequest` interface. An instance of this class is passed to the `HttpServlet.service()` method on a per-request basis.

```
package com.n1guz.ServletEngine;

import javax.servlet.*;
import javax.servlet.http.*;
import java.util.*;
import java.net.*;
import java.io.*;

public class HttpServletRequestImp extends ServletRequestImp
  implements HttpServletRequest
{
  /**
   * Vector of cookie objects associated with the request.
   */

  protected Vector m_vCookies = null;

  /**
   * Method such as get or post.
   */

  protected String m_strMethod = "get";

  /**
   * Universal Resource Identifier.
   */

  protected String m_strRequestURI = null;

  /**
   * Servlet path as it appears within the URL.
   */

  protected String m_strServletPath = null;

  /**
   * Additional optional path information that follows the Servlet
   * path within the URI.
   */

  protected String m_strPathInfo = null;
```

```
/**
 * Absolute path to the servlet.
 */

protected String m_strPathTranslated = null;

/**
 * Name/Value pairs following the '?' in the URL.
 */

protected String m_strQueryString = null;

/**
 * Remote user.
 */

protected String m_strRemoteUser = null;

/**
 * Authentication scheme.
 */

protected String m_strAuthType = null;

/**
 * Reference to the servlet session.
 */

protected HttpSessionImp m_httpSession = null;

/**
 * Requested session ID.
 */

protected String m_requestedSessionID = null;

/**
 * Hashtable of HTTP headers.
 */

protected Hashtable m_hashHeaders = null;
protected HttpSessionContextImp m_sessionContext = null;
protected boolean m_bSessionFromCookie = false;
protected boolean m_bSessionFromParameter = false;

/* *****************************************************************
*******************************************************************
* NON-EXPOSED METHODS:
* The following methods are for internal use and not exposed
* via the JSDK interface.
*******************************************************************
```

```
******************************************************************/

public HttpServletRequestImp(HttpSessionContextImp sessionContext)
{
  m_hashHeaders = new Hashtable();
  m_vCookies = new Vector();
  m_sessionContext = sessionContext;
}

protected void setRemoteAddressAndName(Socket socket)
{
  InetAddress remoteAddress = socket.getInetAddress();

  byte address[] = remoteAddress.getAddress();

  int nLength = address.length;

  m_strRemoteAddr = "";

  for (int nIndex = 0; nIndex < nLength; nIndex++)
  {
    m_strRemoteAddr += String.valueOf((int) address[nIndex]);

    if (nIndex < nLength-1)
    {
      m_strRemoteAddr += ".";
    }
  }

  m_strRemoteHost = remoteAddress.getHostName();
}

public void loadFrom(Socket socket)
  throws IOException
{
  if (socket == null)
  {
    throw new IllegalArgumentException(
        "HTTPRequest.loadFrom(): null socket");
  }

  setRemoteAddressAndName(socket);

  InputStream is = socket.getInputStream();

  InputStreamReader isr = new InputStreamReader(is);
  BufferedReader br = new BufferedReader(isr);

  parseFirstLine(br);
  parseHeaders(br);
```

```
    if (m_strQueryString != null)
    {
      parseOVPIntoParameters(m_strQueryString);
    }

    parseContent(br);
  }

  protected void parseFirstLine(BufferedReader br)
    throws IOException
  {
    // Read the first line.

    String strLine = br.readLine().trim();

    int nPos1 = strLine.indexOf(' ');
    int nPos2 = strLine.indexOf(' ', nPos1+1);

    m_strMethod = strLine.substring(0,nPos1);
    m_strPathInfo = strLine.substring(nPos1+1,nPos2);
    m_strProtocol = strLine.substring(nPos2+1);

    nPos1 = m_strPathInfo.indexOf('?');

    if (nPos1 >= 0)
    {
      String strQueryString = "";
      if (nPos1 > 0)
      {
        m_strQueryString = m_strPathInfo.substring(nPos1+1);
      }

      m_strPathInfo = m_strPathInfo.substring(0,nPos1);
    }

    m_strRequestURI = m_strPathInfo;
  }

  protected void parseHeaders(BufferedReader br)
    throws IOException
  {
    m_hashHeaders = new Hashtable();

    for (;;)
    {
      String strLine = br.readLine().trim();

      if (strLine.length() == 0)
      {
        break;
      }
```

```
int nPos = strLine.indexOf(':');

String strOption =
  strLine.substring(0,nPos).trim().toLowerCase();

String strValue = strLine.substring(nPos+1).trim();

System.out.println(strOption + ": " + strValue);

if (strOption.equalsIgnoreCase("host"))
{
    m_strRemoteHost = strValue;
}
if (strOption.equalsIgnoreCase("content-length"))
{
  m_nContentLength = Integer.parseInt(strValue);
}
else if (strOption.equalsIgnoreCase("content-type"))
{
  m_strContentType = strValue;
}
else if (strOption.equalsIgnoreCase("cookie"))
{
  String strCookie = null;

  while (strValue.length() != 0)
  {
    nPos = strValue.indexOf(';');

    if (nPos == -1)
    {
      strCookie = strValue;
      strValue = "";
    }
    else
    {
      strCookie = strValue.substring(0,nPos);

      if ((nPos+1) == strValue.length())
      {
        strValue = "";
      }
      else
      {
        strValue = strValue.substring(nPos+1);
      }
    }

    nPos = strCookie.indexOf('=');
    String strCookieName = null;
    String strCookieValue = null;
```

```
        if (nPos == -1)
        {
          strCookieName = strCookie.trim();
          strCookieValue = "null";
        }
        else
        {
          strCookieName =
              convert(strCookie.substring(0,nPos).trim());

            strCookieValue =
                convert(strCookie.substring(nPos+1).trim());
        }

        Cookie cookie = null;

        try
        {
          cookie =
              new Cookie(strCookieName,strCookieValue);

          synchronized (m_vCookies)
          {
            m_vCookies.addElement(cookie);
          }
        }
        catch (Exception e)
        {
        }
      }
    }

    m_hashHeaders.put(strOption,strValue);
  }
}

protected void addParameter(String strOption, String strValue)
{
  Object objParameter = m_hashParameters.get(strOption);

  if (objParameter == null)
  {
    m_hashParameters.put(strOption,strValue);
  }
  else
  {
    Vector vParameter = null;

    if (objParameter instanceof String)
    {
      vParameter = new Vector();
```

```
        vParameter.addElement(objParameter);
        m_hashParameters.put(strOption,vParameter);
      }
      else
      {
        vParameter = (Vector) objParameter;
      }

      vParameter.addElement(strValue);
    }
  }

  protected void parseOVPIntoParameters(String strQueryString)
    throws IOException
  {
    for (;;)
    {
      if (strQueryString.length() == 0)
      {
        break;
      }

      int nPos = strQueryString.indexOf('&');

      String strOVP = "";
      if (nPos > 1)
      {
        strOVP = strQueryString.substring(0,nPos);
        if (strQueryString.length()-nPos > 0)
        {
          strQueryString = strQueryString.substring(nPos+1);
        }
      }
      else
      {
        strOVP = strQueryString;
        strQueryString = "";
      }

      nPos = strOVP.indexOf('=');

      if (nPos > 0)
      {
        String strOption = "";
        String strValue = "";

        strOption = convert(strOVP.substring(0,nPos));
        if (strOVP.length()-nPos > 0)
        {
          strValue = convert(strOVP.substring(nPos+1));
        }
```

```
        addParameter(strOption,strValue);
      }
    }
  }

  protected void parseContent(BufferedReader br)
    throws IOException
  {
    if (m_nContentLength > 0)
    {
      char buffer[] = new char[m_nContentLength];
      br.read(buffer,0,m_nContentLength);

      parseOVPIntoParameters(new String(buffer));
    }
  }

  public void setSessionFromCookie(HttpSession session)
  {
    m_bSessionFromCookie = true;
    m_httpSession = (HttpSessionImp) session;
  }

  public void setSessionFromParameter(HttpSession session)
  {
    m_bSessionFromParameter = true;
    m_httpSession = (HttpSessionImp) session;
  }

  /**
   * String convert(String strSource)
   *
   * This method converts a source URL segment into it correct text.
   * When a url is converted for transmission, non alphanumeric
   * characters are represented as a hex value.  The '+' character
   * is used to represent a space.
   *
   * @param strSource- the source string to be coverted.
   * @return the converted text.
   */

  protected String convert(String strSource)
  {
    // Remove whitespace at the begining and ending of the source.

    strSource = strSource.trim();

    // Get the length of the source.

    int nLen = strSource.length();
```

```
    // Look at each character in the source.

    for (int nIndex=0; nIndex < nLen; nIndex++)
    {
      // Get a character.

      char c = strSource.charAt(nIndex);

      // If it is a '+' convert it to a space.

      if (c == '+')
      {
        String strFix = strSource.substring(0,nIndex);
        strFix += ' ';
        if (nIndex < nLen-1)
        {
          strFix += strSource.substring(nIndex+1);
        }

        strSource = strFix;
      }

      // If it is a '%' convert it and the two digits that follow
      // into the character represented by the two digit hex value.

      else if (c == '%')
      {
        String strFix = strSource.substring(0,nIndex);

        String strConversion =
            strSource.substring(nIndex+1, nIndex+3);

        c = (char) Integer.parseInt(strConversion,16);

        strFix += c;

        if (nIndex < nLen-3)
        {
          strFix += strSource.substring(nIndex+3);
        }
        nLen -= 2;

        strSource = strFix;
      }
    }
    return strSource;
  }

  public void setServletPath(String strValue)
  {
```

```
    m_strServletPath = strValue;
  }

  public void setPathInfo(String strValue)
  {
    m_strPathInfo = strValue;
  }

/* *******************************************************************
 *******************************************************************
 * EXPOSED METHODS:
 * The following methods are required and exposed by the JSDK
 * interface.
 *******************************************************************
 *******************************************************************/

  /**
   * Gets the authentication scheme of this request.  Same as the
   * CGI variable AUTH_TYPE.
   *
   * @return this request's authentication scheme, or null if none.
   */

  public String getAuthType()
  {
    return m_strAuthType;
  }

  /**
   * Gets the array of cookies found in this request.
   *
   * @return the array of cookies found in this request
   */

  public Cookie[] getCookies()
  {
    Cookie retvalCookies[] = null;

    synchronized (m_vCookies)
    {
      int nCount = m_vCookies.size();

      retvalCookies = new Cookie[nCount];

      for (int nIndex = 0; nIndex < nCount; nIndex++)
      {
        retvalCookies[nIndex] =
            (Cookie) m_vCookies.elementAt(nIndex);
      }
    }
```

```
    return retvalCookies;
  }

  /**
   * Gets the value of the requested date header field of this
   * request.  If the header can't be converted to a date, the
   * method throws an IllegalArgumentException.  The case of the
   * header field name is ignored.
   *
   * @param name the String containing the name of the requested
   * header field
   * @return the value the requested date header field, or -1 if not
   * found.
   */

  public long getDateHeader(String strName)
  {
    if (strName == null)
    {
      throw new IllegalArgumentException(
          "HTTPRequest.getHeader(): null header name");
    }

    long lRetval = -1;

    strName = strName.toLowerCase();

    String strValue = (String) m_hashHeaders.get(strName);

    if (strValue != null)
    {
      try
      {
        lRetval = Date.parse(strValue);
      }
      catch (Exception e)
      {
        lRetval = -1;
      }
    }

    return lRetval;
  }

  /**
   * Gets the value of the requested header field of this request.
   * The case of the header field name is ignored.
   *
   * @param name the String containing the name of the requested
   * header field
```

```
 * @return the value of the requested header field, or null if not
 * known.
 */

public String getHeader(String strName)
{
  if (strName == null)
  {
    throw new IllegalArgumentException(
        "HTTPRequest.getHeader(): null header name");
  }

  String strRetval = null;

  strName = strName.toLowerCase();

  Object obj = m_hashHeaders.get(strName);

  if (obj != null)
  {
    strRetval = obj.toString();
  }

  return strRetval;
}

/**
 * Gets the header names for this request.
 *
 * @return an enumeration of strings representing the header names
 * for this request. Some server implementations do not allow
 * headers to be accessed in this way, in which case this method
 * will return null.
 */

public Enumeration getHeaderNames()
{
  return m_hashHeaders.keys();
}

/**
 * Gets the value of the specified integer header field of this
 * request.  The case of the header field name is ignored.  If the
 * header can't be converted to an integer, the method throws a
 * NumberFormatException.
 *
 * @param name the String containing the name of the requested
 * header field
 * @return the value of the requested header field, or -1 if not
 * found.
 */
```

```
public int getIntHeader(String strName)
{
  if (strName == null)
  {
    throw new IllegalArgumentException(
        "HTTPRequest.getHeader(): null header name");
  }

  int nRetval = -1;

  strName = strName.toLowerCase();

  String strValue = (String) m_hashHeaders.get(strName);

  if (strValue != null)
  {
    try
    {
      nRetval = Integer.parseInt(strValue);
    }
    catch (Exception e)
    {
      nRetval = -1;
    }
  }

  return nRetval;
}

/**
 * Gets the HTTP method (for example, GET, POST, PUT) with which
 * this request was made. Same as the CGI variable REQUEST_METHOD.
 *
 * @return the HTTP method with which this request was made
 */

public String getMethod()
{
  return m_strMethod;
}

/**
 * Gets any optional extra path information following the servlet
 * path of this request's URI, but immediately preceding its query
 * string. Same as the CGI variable PATH_INFO.
 *
 * @return the optional path information following the servlet
 * path, but before the query string, in this request's URI; null
 * if this request's URI contains no extra path information
 */
```

```
public String getPathInfo()
{
  return m_strPathInfo;
}

/**
 * Gets any optional extra path information following the servlet
 * path of this request's URI, but immediately preceding its query
 * string, and translates it to a real path.  Similar to the CGI
 * variable PATH_TRANSLATED
 *
 * @return extra path information translated to a real path or
 * null if no extra path information is in the request's URI
 */

public String getPathTranslated()
{
  return m_strPathTranslated;
}

/**
 * Gets any query string that is part of the HTTP request URI.
 * Same as the CGI variable QUERY_STRING.
 *
 * @return query string that is part of this request's URI, or
 * null if it contains no query string
 */

public String getQueryString()
{
  return m_strQueryString;
}

/**
 * Gets the name of the user making this request.  The user name
 * is set with HTTP authentication.  Whether the user name will
 * continue to be sent with each subsequent communication is
 * browser-dependent.  Same as the CGI variable REMOTE_USER.
 *
 * @return the name of the user making this request, or null if
 * not known.
 */

public String getRemoteUser()
{
  return m_strRemoteUser;
}

/**
 * Gets the session id specified with this request.  This may
 * differ from the actual session id.  For example, if the request
```

```
     * specified an id for an invalid session, then this will get a
     * new session with a new id.
     *
     * @return the session id specified by this request, or null if
     * the request did not specify a session id
     *
     * @see #isRequestedSessionIdValid
     */

   public String getRequestedSessionId ()
   {
     return m_requestedSessionID;
   }

   /**
    * Gets, from the first line of the HTTP request, the part of this
    * request's URI that is to the left of any query string.
    * For example,
    *
    * <blockquote>
    * <table>
    * <tr align=left><th>First line of HTTP request<th>
    * <th>Return from <code>getRequestURI</code>
    * <tr><td>POST /some/path.html HTTP/1.1<td><td>/some/path.html
    * <tr><td>GET http://foo.bar/a.html HTTP/1.0
    * <td><td>http://foo.bar/a.html
    * <tr><td>HEAD /xyz?a=b HTTP/1.1<td><td>/xyz
    * </table>
    * </blockquote>
    *
    * <p>To reconstruct a URL with a URL scheme and host, use the
    * method javax.servlet.http.HttpUtils.getRequestURL, which returns
    * a StringBuffer.
    *
    * @return this request's URI
    * @see javax.servlet.http.HttpUtils#getRequestURL
    */

  public String getRequestURI()
  {
    return m_strRequestURI;
  }

  /**
   * Gets the part of this request's URI that refers to the servlet
   * being invoked. Analogous to the CGI variable SCRIPT_NAME.
   *
   * @return the servlet being invoked, as contained in this
   * request's URI
   */
```

```
public String getServletPath()
{
  return m_strServletPath;
}

/**
 * Gets the current valid session associated with this request, if
 * create is false or, if necessary, creates a new session for the
 * request, if create is true.
 *
 * Note: to ensure the session is properly maintained,
 * the servlet developer must call this method (at least once)
 * before any output is written to the response.
 *
 * Additionally, application-writers need to be aware that newly
 * created sessions (that is, sessions for which
 * HttpSession.isNew returns true) do not have any
 * application-specific state.
 *
 * @return the session associated with this request or null if
 * create was false and no valid session is associated
 * with this request.
 */

public HttpSession getSession ()
{
  return getSession(true);
}

public HttpSession getSession (boolean bCreate)
{
  if (((m_httpSession == null) || !m_httpSession.isValid()) &&
    bCreate)
  {
    m_httpSession = new HttpSessionImp(m_sessionContext);
    m_sessionContext.put(m_httpSession.getId(), m_httpSession);
  }

  return m_httpSession;
}

/**
 * Checks whether the session id specified by this request came in
 * as a cookie.  (The requested session may not be one returned by
 * the <code>getSession</code> method.)
 *
 * @return true if the session id specified by this request came
 * in as a cookie; false otherwise
 *
```

```
     * @see #getSession
     */

    public boolean isRequestedSessionIdFromCookie ()
    {
      return m_bSessionFromCookie;
    }

    /**
     * Checks whether the session id specified by this request came in
     * as part of the URL.  (The requested session may not be the one
     * returned by the <code>getSession</code> method.)
     *
     * @return true if the session id specified by the request for this
     * session came in as part of the URL; false otherwise
     *
     * @see #getSession
     */

    public boolean isRequestedSessionIdFromURL ()
    {
      return m_bSessionFromParameter;
    }

    public boolean isRequestedSessionIdFromUrl ()
    {
      return m_bSessionFromParameter;
    }

    /**
     * Checks whether this request is associated with a session that
     * is valid in the current session context.  If it is not valid,
     * the requested session will never be returned from the
     * <code>getSession</code> method.
     *
     * @return true if this request is assocated with a session that
     * is valid in the current session context.
     *
     * @see #getRequestedSessionId
     * @see javax.servlet.http.HttpSessionContext
     * @see #getSession
     */

    public boolean isRequestedSessionIdValid ()
    {
      return m_httpSession.isValid();
    }
}
```

The HttpServletResponseImp Class

The `HttpServletResponseImp` class is used as the object that implements the `HttpServletResponse` interface. An instance of this class is passed to the `HttpServlet.service()` method on a per-request basis.

```
package com.n1guz.ServletEngine;

import javax.servlet.*;
import javax.servlet.http.*;
import java.util.*;
import java.net.*;
import java.io.*;

public class HttpServletResponseImp
    extends ServletResponseImp implements HttpServletResponse
{
  protected Vector m_vCookies = new Vector();
  protected Hashtable m_hashFields = new Hashtable(31);
  protected int m_nStatusCode = -1;
  protected String m_strStatusMessage = null;
  protected long m_lDate = -1;
  protected String m_strBodyMessage = null;
  protected HttpServletRequestImp m_request = null;

  /* *****************************************************************
   *****************************************************************
   * NON-EXPOSED METHODS:
   * The following methods are for internal use and not exposed
   * via the JSDK interface.
   *****************************************************************
   *****************************************************************/

  public HttpServletResponseImp(HttpServletRequestImp request)
  {
    m_request = request;
  }

  public void send(Socket socket)
    throws IOException
  {
    ByteArrayOutputStream baosResponse =
      new ByteArrayOutputStream(2024);
    OutputStreamWriter oswResponse =
      new OutputStreamWriter(baosResponse);
    PrintWriter pw = new PrintWriter(oswResponse);

    pw.println("HTTP/1.1 " + m_nStatusCode + " " +
      m_strStatusMessage);

    // Assume there is not content.
```

```
boolean bHasContent = false;

// For good measure make sure that the content sources have been
// flushed.  If the servlet did a close on their stream, we might
// catch an IOException, so be prepaired.

try
{
  if (m_bosContent != null)
  {
    m_bosContent.flush();
  }

  if (m_pwContent != null)
  {
    m_pwContent.flush();
    m_oswContent.flush();
  }
}
catch (IOException e)
{
}
// See if the servlet wrote some content.

m_nContentLength = m_baosContent.size();

// If the servlet didn't write content, see if the set status
// produced a body message.

if ((m_nContentLength == 0) && (m_strBodyMessage != null))
{
  OutputStreamWriter oswContent =
    new OutputStreamWriter(m_baosContent);
  PrintWriter pwContent = new PrintWriter(oswContent);

  pwContent.println(m_strBodyMessage);
  pwContent.flush();
  oswContent.flush();
}

// Check the content length again in case there was a body
// message.

m_nContentLength = m_baosContent.size();

// Load up content related headers.

if (m_nContentLength > 0)
{
  bHasContent = true;
  setIntHeader("Content-Length", m_nContentLength);
```

```
    setHeader("Content-Type", m_strContentType);
  }

  // Print out headers.

  Enumeration enum = m_hashFields.keys();

  while (enum.hasMoreElements())
  {
    String strOption = (String) enum.nextElement();
    Object objValue = m_hashFields.get(strOption);

    pw.print(strOption + ": ");

    if (objValue instanceof Date)
    {
      pw.println(objValue.toString());
    }
    else
    {
      pw.println(objValue.toString());
    }
  }

  enum = m_vCookies.elements();

  while (enum.hasMoreElements())
  {
    Cookie cookie = (Cookie) enum.nextElement();

    String strCookie = cookie.getName();
    String strCookieValue = cookie.getValue();
    int nMaxAge = cookie.getMaxAge();
    String strPath = cookie.getPath();
    String strDomain = cookie.getDomain();
    String strComment = cookie.getComment();
    String strVersion = String.valueOf(cookie.getVersion());

    String strExpDate = null;

    if (nMaxAge != -1)
    {
      GregorianCalendar gc = new GregorianCalendar();
      gc.add(gc.SECOND,nMaxAge);

      strExpDate = getCookieTime(gc);
    }

    pw.print("Set-Cookie: ");
    pw.print(strCookie + "=" + strCookieValue + ";");
```

```
        if (nMaxAge != -1)
        {
          pw.print(" expires=" + strExpDate + ";");
        }
        if (strPath != null)
        {
          pw.println(" path=" + strPath + ";");
        }
        if (strDomain != null)
        {
          pw.println(" domain=" + strDomain + ";");
        }
        if (strVersion != null)
        {
          pw.println(" version=" + strVersion + ";");
        }
        if (strComment != null)
        {
          pw.println(" comment=" + strComment + ";");
        }
      }

      // Terminate the header.

      pw.println();

      // Flush the header.
      pw.flush();
      oswResponse.flush();

      if (bHasContent)
      {
        // Write the content.
        baosResponse.write(m_baosContent.toByteArray());
      }

      OutputStream os = socket.getOutputStream();
      BufferedOutputStream bos = new BufferedOutputStream(os);
      bos.write(baosResponse.toByteArray());
      bos.flush();
      os.flush();
    }

    protected String getCookieTime(GregorianCalendar gc)
    {
      String strRetval = null;

      // Get the day of the week.

      int nDayOfWeek = gc.get(gc.DAY_OF_WEEK);
```

```
String strDayOfWeek = null;
switch (nDayOfWeek)
{
  case GregorianCalendar.SUNDAY:
    strDayOfWeek = "Sunday";
    break;
  case GregorianCalendar.MONDAY:
    strDayOfWeek = "Monday";
    break;
  case GregorianCalendar.TUESDAY:
    strDayOfWeek = "Tuesday";
    break;
  case GregorianCalendar.WEDNESDAY:
    strDayOfWeek = "Wednesday";
    break;
  case GregorianCalendar.THURSDAY:
    strDayOfWeek = "Thursday";
    break;
  case GregorianCalendar.FRIDAY:
    strDayOfWeek = "Friday";
    break;
  case GregorianCalendar.SATURDAY:
    strDayOfWeek = "Saturday";
    break;
}

int nMonth = gc.get(gc.MONTH);
String strMonth = null;

switch (nMonth)
{
  case GregorianCalendar.JANUARY:
    strMonth = "Jan";
    break;
  case GregorianCalendar.FEBRUARY:
    strMonth = "Feb";
    break;
  case GregorianCalendar.MARCH:
    strMonth = "Mar";
    break;
  case GregorianCalendar.APRIL:
    strMonth = "Apr";
    break;
  case GregorianCalendar.MAY:
    strMonth = "May";
    break;
  case GregorianCalendar.JUNE:
    strMonth = "Jun";
    break;
  case GregorianCalendar.JULY:
    strMonth = "Jul";
```

```
        break;
      case GregorianCalendar.AUGUST:
        strMonth = "Aug";
        break;
      case GregorianCalendar.SEPTEMBER:
        strMonth = "Sep";
        break;
      case GregorianCalendar.OCTOBER:
        strMonth = "Oct";
        break;
      case GregorianCalendar.NOVEMBER:
        strMonth = "Nov";
        break;
      case GregorianCalendar.DECEMBER:
        strMonth = "Dec";
        break;
  }

  String strHours =
    String.valueOf(gc.get(GregorianCalendar.HOUR_OF_DAY));

  if (strHours.length() == 1)
  {
    strHours = "0" + strHours;
  }

  String strMinutes =
    String.valueOf(gc.get(GregorianCalendar.MINUTE));

  if (strMinutes.length() == 1)
  {
    strMinutes = "0" + strMinutes;
  }

  String strSeconds =
    String.valueOf(gc.get(GregorianCalendar.SECOND));

  if (strSeconds.length() == 1)
  {
    strSeconds = "0" + strSeconds;
  }

  TimeZone tz = gc.getTimeZone();

  strRetval = strDayOfWeek + ", " +
    gc.get(GregorianCalendar.DAY_OF_MONTH) + "-" +
    strMonth + "-" +
    gc.get(GregorianCalendar.YEAR) + " " +
    strHours + ":" +
    strMinutes + ":" +
    strSeconds + " " +
```

```
      tz.getID();

    return strRetval;
  }

  /* ****************************************************************
   ****************************************************************
   * EXPOSED METHODS:
   * The following methods are required and exposed by the JSDK
   * interface.
   ****************************************************************
   ****************************************************************/

  /**
   * Adds the specified cookie to the response.  It can be called
   * multiple times to set more than one cookie.
   *
   * @param cookie the Cookie to return to the client
   */

  public void addCookie(Cookie cookie)
  {
    m_vCookies.addElement(cookie);
  }

  /**
   * Checks whether the response message header has a field with
   * the specified name.
   *
   * @param name the header field name
   * @return true if the response message header has a field with
   * the specified name; false otherwise
   */

  public boolean containsHeader(String strOption)
  {
    boolean bRetval = false;

    if (null != m_hashFields.get(strOption))
    {
      bRetval = true;
    }

    return bRetval;
  }

  /**
   * Encodes the specified URL for use in the
   * <code>sendRedirect</code> method or, if encoding is not needed,
   * returns the URL unchanged.  The implementation of this method
   * should include the logic to determine whether the session ID
```

```
  * needs to be encoded in the URL.  Because the rules for making
  * this determination differ from those used to decide whether to
  * encode a normal link, this method is seperate from the
  * <code>encodeUrl</code> method.
  *
  * <p>All URLs sent to the HttpServletResponse.sendRedirect
  * method should be run through this method.  Otherwise, URL
  * rewriting canont be used with browsers which do not support
  * cookies.
  *
  * @param url the url to be encoded.
  * @return the encoded URL if encoding is needed; the unchanged URL
  * otherwise.
  *
  * @see #sendRedirect
  * @see #encodeUrl
  */

public String encodeRedirectURL(String url)
{
  return encodeRedirectUrl(url);
}

public String encodeRedirectUrl (String url)
{
  return encodeUrl(url);
}

/**
 * Encodes the specified URL by including the session ID in it,
 * or, if encoding is not needed, returns the URL unchanged.
 * The implementation of this method should include the logic to
 * determine whether the session ID needs to be encoded in the URL.
 * For example, if the browser supports cookies, or session
 * tracking is turned off, URL encoding is unnecessary.
 *
 * <p>All URLs emitted by a Servlet should be run through this
 * method.  Otherwise, URL rewriting cannot be used with browsers
 * which do not support cookies.
 *
 * @param url the url to be encoded.
 * @return the encoded URL if encoding is needed; the unchanged URL
 * otherwise.
 */

public String encodeURL(String url)
{
  return encodeUrl(url);
}

public String encodeUrl (String url)
```

```
{
  HttpSession session = m_request.getSession(false);

  if ((session != null) && (session instanceof HttpSessionImp))
  {
    HttpSessionImp sessionImp = (HttpSessionImp) session;

    if (sessionImp.isValid())
    {
      url += (url.indexOf('?') > -1) ? "&" : "?";

      url += "ServletSessionID=";

      url += session.getId();
    }
  }

  return url;
}

/**
 * Sends an error response to the client using the specified status
 * code and descriptive message.  If setStatus has previously been
 * called, it is reset to the error status code.  The message is
 * sent as the body of an HTML page, which is returned to the user
 * to describe the problem.  The page is sent with a default HTML
 * header; the message is enclosed in simple body tags
 * (&lt;body&gt;&lt;/body&gt;).
 *
 * @param sc the status code
 * @param msg the detail message
 * @exception IOException If an I/O error has occurred.
 */

public void sendError(int nStatusCode, String strMessage)
  throws IOException
{
  setStatus(nStatusCode, strMessage);
}

/**
 * Sends an error response to the client using the specified
 * status code and a default message.
 * @param sc the status code
 * @exception IOException If an I/O error has occurred.
 */

public void sendError(int nStatusCode)
  throws IOException
{
  setStatus(nStatusCode, null);
```

```
}

/**
 * Sends a temporary redirect response to the client using the
 * specified redirect location URL.  The URL must be absolute (for
 * example, <code><em>https://hostname/path/file.html</em></code>).
 * Relative URLs are not permitted here.
 *
 * @param location the redirect location URL
 * @exception IOException If an I/O error has occurred.
 */

public void sendRedirect(String strLocation)
  throws IOException
{
  // Make sure that the print writer and output stream writer are
  // closed.

  if (m_pwContent != null)
  {
      m_pwContent.close();
      m_pwContent = null;
  }

  if (m_oswContent != null)
  {
      m_oswContent.close();
      m_oswContent = null;
  }

  // Set the status for redirect.

  setStatus(HttpServletResponse.SC_MOVED_TEMPORARILY);

  // Create a new print writer for the content.

  m_baosContent = new ByteArrayOutputStream();
  m_oswContent = new OutputStreamWriter(m_baosContent);
  PrintWriter pw = new PrintWriter(m_oswContent);

  // Print the redirect URL.

  pw.println(strLocation);
  pw.close();
}

/**
 *
 * Adds a field to the response header with the given name and
 * date-valued field.  The date is specified in terms of
 * milliseconds since the epoch.  If the date field had already
```

```
 * been set, the new value overwrites the previous one.  The
 * <code>containsHeader</code> method can be used to test for the
 * presence of a header before setting its value.
 *
 * @param name the name of the header field
 * @param value the header field's date value
 *
 * @see #containsHeader
 */

public void setDateHeader(String strOption, long lDate)
{
  m_hashFields.put(strOption, new Date(lDate));
}

/**
 *
 * Adds a field to the response header with the given name and
 * value.  If the field had already been set, the new value
 * overwrites the previous one.  The <code>containsHeader</code>
 * method can be used to test for the presence of a header before
 * setting its value.
 *
 * @param name the name of the header field
 * @param value the header field's value
 *
 * @see #containsHeader
 */

public void setHeader(String strOption, String strValue)
{
  m_hashFields.put(strOption, strValue);
}

/**
 * Adds a field to the response header with the given name and
 * integer value.  If the field had already been set, the new value
 * overwrites the previous one.  The <code>containsHeader</code>
 * method can be used to test for the presence of a header before
 * setting its value.
 *
 * @param name the name of the header field
 * @param value the header field's integer value
 *
 * @see #containsHeader
 */

public void setIntHeader(String strOption, int nValue)
{
  m_hashFields.put(strOption, new Integer(nValue));
}
```

```
/**
 * Sets the status code and message for this response.  If the
 * field had already been set, the new value overwrites the
 * previous one.  The message is sent as the body of an HTML
 * page, which is returned to the user to describe the problem.
 * The page is sent with a default HTML header; the message
 * is enclosed in simple body tags (&lt;body&gt;&lt;/body&gt;).
 *
 * @param sc the status code
 * @param sm the status message
 */

public void setStatus(int nStatusCode, String strMessage)
{
  m_nStatusCode = nStatusCode;
  m_strBodyMessage = strMessage;

  switch (nStatusCode)
  {
    case SC_CONTINUE:
      m_strStatusMessage = "CONTINUE";
      // "Client can continue.";
      break;
    case SC_SWITCHING_PROTOCOLS:
      m_strStatusMessage = "SWITCHING PROTOCOLS";
      // "Switching protocols.";
      break;
    case SC_OK:
      m_strStatusMessage = "OK";
      // "OK";
      break;
    case SC_CREATED:
      m_strStatusMessage = "CREATED";
      // "New resource created on server.";
      break;
    case SC_ACCEPTED:
      m_strStatusMessage = "ACCEPTED";
      // "Request was accepted but was not completed.";
      break;
    case SC_NON_AUTHORITATIVE_INFORMATION:
      m_strStatusMessage = "NON AUTHORITATIVE INFORMATION";
      // "Meta information presented by client did not originate
      // on this server.";
      break;
    case SC_NO_CONTENT:
      m_strStatusMessage = "NO CONTENT";
      // "Request suceeded but no new information to return.";
      break;
    case SC_RESET_CONTENT:
      m_strStatusMessage = "RESET CONTENT";
      // "Agent should reset the document view which caused the
```

```
    // request to be sent.";
    break;
  case SC_PARTIAL_CONTENT:
    m_strStatusMessage = "PARTIAL CONTENT";
    // "Server has fulfilled the partial GET request for the
    // resource.";
    break;
  case SC_MULTIPLE_CHOICES:
    m_strStatusMessage = "MULTIPLE CHOICES";
    strMessage = "(300) Requested resource corresponds to " +
      "any one of a set of representations, each with its " +
      "own specific location.";
    break;
  case SC_MOVED_PERMANENTLY:
    m_strStatusMessage = "MOVED PERMANENTLY";
    strMessage = "(301) Resouce has permanently moved.";
    break;
  case SC_SEE_OTHER:
    m_strStatusMessage = "SEE OTHER";
    strMessage = "(302) Resource has temprarily moved.";
    break;
  case SC_NOT_MODIFIED:
    m_strStatusMessage = "NOT MODIFIED";
    strMessage = "(304) Resource not modified.";
    break;
  case SC_USE_PROXY:
    m_strStatusMessage = "USE PROXY";
    strMessage = "(305) Requested resource MUST be accessed " +
      "through the proxy given by the location field.";
    break;
  case SC_BAD_REQUEST:
    m_strStatusMessage = "BAD REQUEST";
    strMessage = "(400) Request sent by the client is " +
      "syntactically incorrect.";
    break;
  case SC_UNAUTHORIZED:
    m_strStatusMessage = "UNAUTHORIZED";
    strMessage = "(401) Request requires HTTP authentication.";
    break;
  case SC_PAYMENT_REQUIRED:
    m_strStatusMessage = "PAYMENT REQUIRED";
    strMessage = "(402) Reserved for future use.";
    break;
  case SC_FORBIDDEN:
    m_strStatusMessage = "FORBIDDEN";
    strMessage = "(403) Access forbidden.";
    break;
  case SC_NOT_FOUND:
    m_strStatusMessage = "NOT FOUND";
    strMessage = "(404) Requested resource is not available.";
    break;
```

```
case SC_METHOD_NOT_ALLOWED:
  m_strStatusMessage = "METHOD NOT ALLOWED";
  strMessage = "(405) Method specified is not allowed for " +
    "that resource.";
  break;
case SC_NOT_ACCEPTABLE:
  m_strStatusMessage = "NOT ACCEPTABLE";
  strMessage = "(406) Requested resource has " +
    "characteristics unacceptable to client.";
  break;
case SC_PROXY_AUTHENTICATION_REQUIRED:
  m_strStatusMessage = "PROXY AUTHENTICATION REQUIRED";
  strMessage = "(407) Client must first authenticate itself " +
    "with the proxy.";
  break;
case SC_REQUEST_TIMEOUT:
  m_strStatusMessage = "REQUEST TIMEOUT";
  strMessage = "(408) Client did not produce a request " +
    "within the time that the server was prepared to wait.";
  break;
case SC_CONFLICT:
  m_strStatusMessage = "CONFLICT";
  strMessage = "(409) Request could not be completed due " +
    "to a conflict with the current state of the resource.";
  break;
case SC_GONE:
  m_strStatusMessage = "GONE";
  strMessage = "(410) The resource is no longer available " +
    "at the server and no forwarding address is known.";
  break;
case SC_LENGTH_REQUIRED:
  m_strStatusMessage = "LENGTH REQUIRED";
  strMessage = "(411) Request cannot be handled without a " +
    "Content-Length.";
  break;
case SC_PRECONDITION_FAILED:
  m_strStatusMessage = "PRECONDITION FAILED";
  strMessage = "(412) One or more preconditions in the " +
    "request-header evaluated to false.";
  break;
case SC_REQUEST_ENTITY_TOO_LARGE:
  m_strStatusMessage = "REQUEST ENTITY TOO LARGE";
  strMessage = "(413) Request entity too large.";
  break;
case SC_REQUEST_URI_TOO_LONG:
  m_strStatusMessage = "REQUEST URI TOO LONG";
  strMessage = "(414) Request URI too long.";
  break;
case SC_UNSUPPORTED_MEDIA_TYPE:
  m_strStatusMessage = "UNSUPPORTED MEDIA TYPE";
  strMessage = "(415) Format unsupported.";
```

```
      break;
    case SC_INTERNAL_SERVER_ERROR:
      m_strStatusMessage = "INTERNAL SERVER ERROR";
      strMessage = "(500) Internal server error.";
      break;
    case SC_NOT_IMPLEMENTED:
      m_strStatusMessage = "NOT IMPLEMENTED";
      strMessage = "(501) Server does not support the " +
        "functionality needed to fulfill the request.";
      break;
    case SC_BAD_GATEWAY:
      m_strStatusMessage = "BAD GATEWAY";
      strMessage = "(502) Server received an invalid response " +
        "from a server it consulted when acting as a proxy or " +
        "gateway.";
      break;
    case SC_SERVICE_UNAVAILABLE:
      m_strStatusMessage = "SERVICE UNAVAILABLE";
      strMessage = "(503) Server is temprarily overloaded and " +
        "unable to handle the request.";
      break;
    case SC_GATEWAY_TIMEOUT:
      m_strStatusMessage = "GATEWAY TIMEOUT";
      strMessage = "(504) Server did not receive a timely " +
        "response from an upstream server while acting as " +
        "a gateway or proxy.";
      break;
    case SC_HTTP_VERSION_NOT_SUPPORTED:
      m_strStatusMessage = "HTTP VERSION NOT SUPPORTED";
      strMessage = "(505) Server does not support specified " +
        "HTTP version.";
      break;
  }

  if (m_strBodyMessage == null)
  {
    m_strBodyMessage = strMessage;
  }
}

/**
 * Sets the status code for this response.  This method is used to
 * set the return status code when there is no error (for example,
 * for the status codes SC_OK or SC_MOVED_TEMPORARILY).  If there
 * is an error, the <code>sendError</code> method should be used
 * instead.
 *
 * @param sc the status code
 *
 * @see #sendError
 */
```

```
    public void setStatus(int nStatusCode)
    {
      setStatus(nStatusCode,null);
    }
  }
```

The MimeServlet Class

Finally, the `MimeServlet` class is a servlet that serves generic Multi-Purpose Internet Mail Extensions (MIME) content. Most of the time, you will use an HTTP server to which the manufacturer has added a Java-based servlet engine. I chose to take a slightly backward approach in my implementation, however. Rather than developing an HTTP server, I developed a servlet engine. I then added the following MIME servlet for accessing MIME content. This servlet simply loads a file into a buffer send it in the content of the response.

```
package com.n1guz.ServletEngine;

import java.io.*;
import java.util.*;
import javax.servlet.*;
import javax.servlet.http.*;

public class MimeServlet extends HttpServlet
{
  protected static int sm_nBufferSize = 1024;
  public void init(ServletConfig config)
  throws ServletException
  {
    // Always pass the ServletConfig object to the super class.

    super.init(config);
  }

  protected String sm_strBasePath = null;

  public void doGet(HttpServletRequest request,
    HttpServletResponse response)
    throws ServletException, IOException
  {
    if (sm_strBasePath == null)
    {
      File fileBasePath = new File(".");
      sm_strBasePath = fileBasePath.getAbsolutePath();
      int nLen = sm_strBasePath.length();
      sm_strBasePath = sm_strBasePath.substring(0,nLen-2);
    }

    // Get the translated path.
```

```
File file = new File(sm_strBasePath + request.getPathInfo());
String strPathTranslated = file.getAbsolutePath();

ServletOutputStream os = response.getOutputStream();

int nPos = strPathTranslated.lastIndexOf('.');

String strExt = "txt";

if (nPos > 0)
{
  strExt = strPathTranslated.substring(nPos+1);
}

ServletConfig servletConfig = getServletConfig();
ServletContext servletContext =
  servletConfig.getServletContext();

String strContentType = servletContext.getMimeType(strExt);

if (strContentType == null)
{
  strContentType = "text/html";
}

response.setContentType(strContentType);

if (file.exists())
{
  try
  {
    FileInputStream fis = new FileInputStream(file);
    byte buffer[] = new byte[sm_nBufferSize];

    for(;;)
    {
      int nAvailable = fis.available();

      if (nAvailable > sm_nBufferSize)
      {
        nAvailable = sm_nBufferSize;
      }

      if (nAvailable > 0)
      {
        fis.read(buffer,0,nAvailable);
        os.write(buffer,0,nAvailable);
      }
      else
      {
        break;
```

```
          }
        }
        fis.close();
      }
      catch (IOException e)
      {
      }
    }

    os.close();
  }
}
```

APPENDIX D

Companion CD-ROM

What's on the Companion CD-ROM?

The companion CD-ROM contains three directories: /code, /SDKs, and /Sun. The /code directory contains the software examples shown in the book. The /SDKs directory contains 3rd party software development kits that are useful for developing WAP Servlets. The /Sun directory contains version 1.3 of Sun's Java 2 Software Development Kit (SDK), version 2.1 of Sun's Java Servlet Developers Kit (JSDK) and version 1.0 of Sun's Forte4j.

NOTE

Please refer to the _readme.txt file on the CD-ROM for late breaking information regarding its content.

Hardware Requirements

For Microsoft Windows, Linux, and Macintosh systems, I would recommend the following hardware configuration as a minimum:

- 300 MHz Processor (500 MHz preferred)
- 128 MB of RAM (256 MB preferred)
- At least 100 MB of available disk space
- CD-ROM Drive

Installing the Software

The following section describes the installation of software components provided on the CD-ROM.

Forte4j 1.0

If you do not already have an Integrated Development Environment (IDE) for Java development, I would highly recommend using Sun Microsystems' Forte4j. This IDE is brought to you by the source of Java technology (Sun Microsystems), and it is a Pure Java application. While Forte4j is a little resource intensive and has some minor quirks, it is certainly moving in the right direction. Furthermore, there is a lot to be said for an IDE that is exactly the same on your operating system of choice. I have personally used Forte4j on MS Windows 2000 and Linux (on both Intel [Redhat] and Apple's iMac [LinuxPPC2000]).

To install the software, follow these simple steps:

1. Start your operating system of choice.
2. Place the CD-ROM into your CD-ROM drive.
3. Start a command shell (or terminal window).
4. Mount the CD-ROM if required by the operating system.
5. Navigate to the `/Sun` directory on the CD-ROM.
6. Execute the installation application that is appropriate for your operating system. (Note: if you already have a version of the Java Runtime Environment (JRE) installed, you may simply use the class version of the installation.)
7. Follow the instructions that appear on the screen.

Make sure that you agree with the terms of Sun's license before using this software. The install image is provided on the CD-ROM to assist users that are limited to slow Internet connections. You should visit the Sun's site (www.javasoft.com) and register as a user prior to using the software.

Java 2 SDK 1.3

If you do not already have a Java Development Kit (JDK) on your system, I would highly recommend using Sun Microsystems' Java 2 SDK version 1.3. Using this environment will help insure that you are developing Pure Java servlets that will have the best possibility of running on a wide range of operating systems and servlet engines. The Java 2 SDK 1.3 is installed using the Forte4j installation (see Forte4j 1.0 below).

During the installation of Forte4j, you will be prompted for the location to install the JDK.

Make sure that you agree with the terms of Sun's license before using this software. The install image is provided on the CD-ROM to assist users that are limited to slow

Internet connections. You should visit the Sun's site and register as a user prior to using the software.

Java Servlet Development Kit (JSDK) 2.1

If you do not already have the Java Servlet Development Kit (JSDK) 2.1 installed on your system, you can install it from the CD-ROM. The JSDK 2.1 is installed using the Forte4j installation (see Forte4j 1.0).

Once Forte4j 1.0 is installed on your system, you will find the JSDK at `/forte4j/lib/ext/servlets.jar` where `/forte4j` is the directory where you installed Forte4j.

Make sure that you agree with the terms of Sun's license before using this software. The install image is provided on the CD-ROM to assist users that are limited to slow Internet connections. You should visit the Sun's site (www.javasoft.com) and register as a user prior to using the software.

Phone.Com SDK 4.0

The Phone.Com SDK provides a suite of tools that is a must have for any WAP Servlet developer. These tools work on MS-Windows based systems. I have personal used the SDK on Windows 95, Windows NT, and Windows 2000.

To install the software, follow these simple steps:

1. Start Windows on your computer.
2. Place the CD-ROM into your CD-ROM drive.
3. Using the explorer, navigate to the `/SDKs/Phone` directory on the CD-ROM.
4. Double click the `upsdkW30e.exe` program.
5. Follow the instructions that appear on the screen.

Make sure that you agree with the terms of Phone.Com's license before using this software. The install image is provided on the CD-ROM to assist users that are limited to slow Internet connections. You should visit the Phone.Com site and register as a user prior to using the software.

Nokia SDK

As I am writing this Appendix, I do not have a final copy of the Nokia SDK. I hope to have a version of the SDK on the CD-ROM, however. If I am unable to get a copy by the time the CD-ROM must go to press, I urge you to navigate to Nokia's site, register as a developer and download their SDK.

To install the software (if present), follow these simple steps:

1. Start Windows on your computer.
2. Place the CD-ROM into your CD-ROM drive.
3. Using the explorer, navigate to the `/SDKs/Nokia` directory on the CD-ROM.
4. Double click the setup program icon.
5. Follow the instructions that appear on the screen.

NOTE

As stated throughout the book, it is important to use multiple WAP simulators if you want your application to look good on more than one handset.

Make sure that you agree with the terms of Nokia's license before using this software. The install image is provided on the CD-ROM to assist users that are limited to slow Internet connections. You should visit the Nokia's (www.nokia.com) site and register as a user prior to using the software.

Example Source Code

The source code for the examples in the book are provided in the `/code` directory.

To install the software, follow these simple steps:

1. Start your operating system of choice.
2. Place the CD-ROM into your CD-ROM drive.
3. Mount the CD-ROM if required by the operating system.
4. Navigate to the / directory on the CD-ROM.
5. Recursively copy the `/code` directory to your hard drive.

NOTE

If you are using or plan to use Forte4j, I would recommend creating a directory called `ForteFilesystems`. In a Linux environment, I would recommend creating this directory in your root (/) directory. In a MS-Windows environment, I would recommend creating this directory in the root directory of your drive of choice. When copying the /code directory, copy it to a directory under `ForteFilesystems`.

Using the Software

The following section describes the use of software components provided on the CD-ROM.

Forte4j 1.0

When you run Forte4j 1.0 the first time, it will want to look for updates to the software. Therefore, make sure that you have an active Internet connection at the time you first run the IDE: there will be updates available. Accept all of the updates by following the on screen directions.

Like most IDE's Forte4j has a notion of projects; however, the project is a little different from what you might have seen in other environments. In Forte4j, the project is primarily a collection of file systems and properties. A file system is essentially a jar (or zip file) that contains classes that are rooted in the default package or directories on the local file system that are also rooted in the default package.

When you initially start Forte4j, you will be drop into a default project with a default file system. I recommend creating a new project for your WAP servlet work. To do this, select `Project >> New Project` from the main menu. In the dialog that appears, type *WAPServlets*, and click *OK*. When asked if you want to start with a new file system, click *New*. When asked if you want to save the current project, select *Yes*. Next, navigate to the `ForteFilesystems` directory you created above. Then select the `code` directory and click `Mount`.

At this point you can click on the `Project` tab in the `Explorer` window. Since the project is new, it will be empty. By right mouse clicking on the project in the `Explorer` window and selecting add existing, you can navigate through the mounted filesystem(s) and add folders to the project. Use this technique to add the following folders to your project:

- /book/chap02/HelloWorld
- /book/chap04/DumpServlet
- /book/chap05/PageViewer
- /book/chap06/Grocery
- /com/n1guz/ServletEngine

Right mouse clicking on each folder and selecting *Build All* will generate the associated class files. Expanding the `ServletEngine` folder, right mouse clicking on the class `ServletEngine`, and selecting `execute` will cause the servlet engine to be executed.

IMPORTANT

The servlets and the servlet engine expect the code directory to be the default working directory. Therefore you need to establish that directory as the working directory of your project. To do this select `Project >> Settings` from the main menu. Next, navigate down the project settings explorer as follows `Execution Types >> External Execution >> External Execution`. Next, select the Expert tab, click on the value of the Working Directory, and navigate to the `ForteFilesystems` directory you created above and select the `code` directory.

You are now ready to select a WAP SDK and navigate to a servlet.

Servlet Engine

The servlet engine is executed by right mouse clicking on the `ServletEngine` class in Forte4j and selecting `Execute`. The behavior of the servlet engine is guided by the file

ServletEngine.conf located in the ServletEngine directory. This section describes the content of the ServletEngine.conf file. By editing this file, you can change the way the servlet engine works.

listen-port

The listen-port property is used to establish the TCP port on which the servlet engine listens for incoming HTTP requests. The default value is 80.

```
com.n1guz.ServletEngine.listen-port=<tcp-port>
```

backlog

The backlog property is used to establish the backlog of TCP connections that will be accepted before processing of the connection commences. The default value is 10.

```
com.n1guz.ServletEngine.backlog=<number>
```

mime-type

Several mime-type properties are used to establish mime-types associated with file extensions.

```
com.n1guz.ServletEngine.mime-type.<mime-type>=<ext>[,<ext>]
```

servlet

The servlet property is used to establish the virtual path (relative to the document root) of a given servlet. Each servlet property represents a single servlet.

```
com.n1guz.ServletEngine.servlet.<rooted path>=<class path>
```

init-parameter

The init-parameter is used to establish initialization parameters common to all servlets. These parameters may be accessed using the ServletConfig.getInitParameter() method.

```
com.n1guz.ServletEngine.init-parameter.<name>=<value>
```

Phone.Com UP.SDK 4.0

To execute the Phone.Com SDK 4.0 UP.Simulator, click on Start >> Programs >> UP.SDK 4.0 >> UP.Simulator. Once both the ServletEngine (as previously described) and the UP.Simulator are running, you can enter a URL into the combo box labeled Go on the UP.Simulator. For example, enter http://localhost/

PageViewer and press return. If everything has been configured correctly, you will see main screen of the PageViewer servlet as described and shown in Chapter 5.

NOTE

When the UP.Simulator is executing, you may select keys on the phone to navigate and enter values: as a user would if they were using the actual handset. However, Phone.Com has made the process of development a little simpler by also allowing you to enter data via the keyboard. Characters and numbers may be entered by typing them on the keyboard. Furthermore, a few useful keyboard shortcuts are also recognized: the *Enter* key can be used for the *accept* soft-key. The *ESC* key can be used for the *back* softkey. The arrow keys may be used for their handset equivalents.

Refer to the documentation that came with the UP.Simulator for more detailed information.

User Assistance and Information

The software accompanying this book is being provided as is without warranty or support of any kind. Should you require basic installation assistance, or if your media is defective, please call our product support number at (212) 850-6194 weekdays between 9 A.M. and 4 P.M. Eastern Standard Time. Or, we can be reached via e-mail at: wprtusw@jwiley.com.

To place additional orders or to request information about other Wiley products, please call (800) 879-4539.

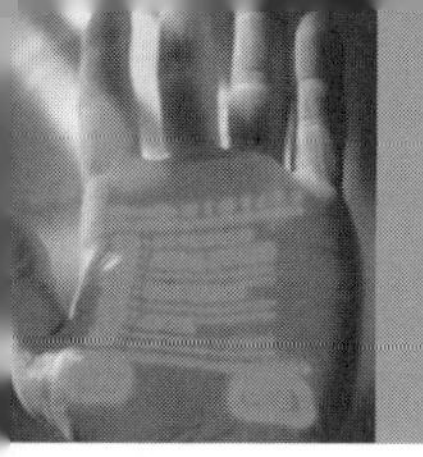

Bibliography

Comer, D. E., 1995, *Internetworking with TCP/IP*, Third Edition, Prentice Hall, Upper Saddle River, NJ.

Flanagan, D., 1997, *Java in a Nutshell*, Second Edition, O'Reilly & Associates, Inc., Sebastopol, CA.

Goodwill, J., 1999, *Developing Java Servlets: The Authoritative Solution*, Howard Sams Publishing, Indianapolis, IN.

Hunter, J. et. al., 1998, *Java Servlet Programming*, O'Reilly & Associates, Inc., Sebastopol, CA.

Mann, S., 1999, *Programming Applications with the Wireless Application Protocol: The Complete Developer's Guide*, John Wiley & Sons, Inc., New York, NY.

Mullen, R., 1998, *The HTML 4 Programmer's Reference*, Ventana Communications Group, Research Triangle Park, NC.

Shannon, B. et. al., 2000, *Java 2 Platform, Enterprise Edition*, Addison-Wesley, Reading, MA.

Stevens, W. R., 1990, *Unix Network Programming*, Prentice Hall, Englewood Cliffs, NJ.

Wireless Application Protocol Forum, Ltd., 1999, *Official Wireless Application Protocol*, John Wiley & Sons, Inc., New York, NY.

[RFC 768] Postel, J. *User Datagram Protocol.* IETF Request for Comment 768.

[RFC 791] Postel, J. *Internet Protocol.* IETF Request for Comment 791.

[RFC 1436] Anklesaria, F. et al. *The Internet Gopher Protocol.* IETF Request for Comment 1436.

[RFC 1866] Berners-Lee, T. et al. *Hypertext Markup Language.* IETF Request for Comment 1866.

[RFC 2045] Freed, N. et al. *Multipurpose Internet Mail Extensions.* IETF Request for Comment 2045.

[RFC 2109] Kristol, D. et al. *HTTP State Management Mechanism.* IETF Request for Comment 2109.

[RFC 2246] Dierks, T. et al. *The TLS Protocol Version 1.0.* IETF Request for Comment 2246.

[RFC 2616] Fielding, R. et al. *Hypertext Transfer Protocol.* IETF Request for Comment 2616.

[RFC 7930] Postel, J. *Transmission Control Protocol.* IETF Request for Comment 7930.

[WAPPAP] Wireless Application Protocol Forum, Ltd., 1999, *Push Access Protocol Specification*, www.wapforum.org/.

[WAPPMS] Wireless Application Protocol Forum, Ltd., 1999, *Push Message Specification*, www.wapforum.org/.

[WAPPPGSS] Wireless Application Protocol Forum, Ltd., 1999, *Push Proxy Gateway Service Specification*, www.wapforum.org/.

[WAPWDP] Wireless Application Protocol Forum, Ltd., 1999, *Wireless Datagram Protocol*, www.wapforum.org/.

[WAPWSPS] Wireless Application Protocol Forum, Ltd., 1999, *Wireless Session Protocol Specification*, www.wapforum.org/.

[WAPWTLS] Wireless Application Protocol Forum, Ltd., 1999, *Wireless Transport Layer Security*, www.wapforum.org/.

[WAPWTPS] Wireless Application Protocol Forum, Ltd., 1999, *Wireless Transaction Protocol Specification*, www.wapforum.org/.

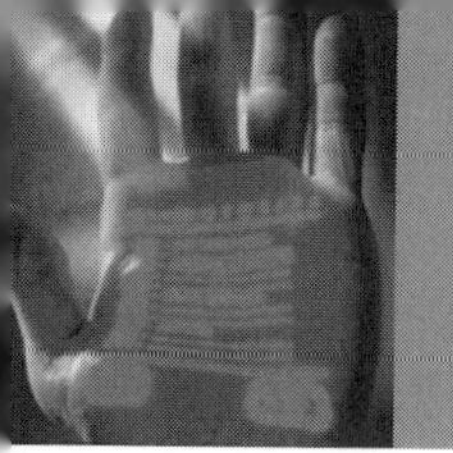

Index

D

E

F

G

H

X

CUSTOMER NOTE: IF THIS BOOK IS ACCOMPANIED BY SOFTWARE, PLEASE READ THE FOLLOWING BEFORE OPENING THE PACKAGE.